Revisiting EU Policy Options for Tackling Climate Change

A Social Cost-Benefit Analysis of GHG Emissions Reduction Strategies

C. Egenhofer

J.C. Jansen

S.J.A. Bakker

and

J. Jussila Hammes

The Centre for European Policy Studies (CEPS) is an independent policy research institute based in Brussels. Its mission is to produce sound analytical research leading to constructive solutions to the challenges facing Europe today. The views expressed in this report are those of the authors writing in a personal capacity and do not necessarily reflect those of CEPS or any other institution with which the authors are associated.

Cover photo shows a low-pressure system swirling off the southwestern coast of Iceland taken by NASA, 4 September 2003.

ISBN 978-92-9079-631-2

Centre for European Policy Studies
Place du Congrès 1, B-1000 Brussels
Tel: 32 (0) 2 229.39.11 Fax: 32 (0) 2 219.41.51
e-mail: info@ceps.be
internet: http://www.ceps.be

ACKNOWLEDGEMENTS

This report has been produced jointly by the Centre for European Policy Studies (CEPS) and the Energy research Centre of the Netherlands (ECN), with support from the Federation of European Rigid Polyurethane Foam Associations (BING), the European Manufacturers of Expanded Polystyrene (EUMEPS), the European Insulation Manufacturers Association (EURIMA) and the European Extruded Polystyrene Insulation Board Association (EXIBA).

A draft of the study was peer-reviewed by Professor Thomas Sterner of the University of Göteborg and Dr. Felix Matthes of Öko Institute, Berlin. The draft report was also discussed during a one-day meeting with some 30 stakeholders. Comments by the peer-reviewers and the stakeholders have been gratefully incorporated.

With the usual disclaimer we acknowledge insightful advice provided by:

- Rick Bradley (International Energy Agency), Randall Bowie (DG TREN, European Commission), Stefan Thomas (Wuppertal Institute), Simon Schmitz (World Business Council for Sustainable Development), Anders Ulf Claussen (Rockwool International), Lena Esteves (EURIMA), Françoise Maon (EUMEPS), Helle Juhler-Kristoffersen (Dansk Industri), Bert de Wel (MiNa-Raad of Flanders) and Matthias Duwe (Climate Action Network)

- CEPS: David Kernohan and Louise van Schaik

- ECN: Luuk Beurskens, Jos Bruggink, Bert Daniëls, Paul Lako, Wietze Lise, Martin Scheepers, Ad Seebregts, Jos Sijm, Koen Smekens, Bas Wetzelaer and Bob van der Zwaan (co-reader).

Contents

List of Tables

List of Figures

PREFACE

Tackling climate change poses one of the world's greatest challenges. The increased need for and use of energy from fossil fuels, and other human activities contribute to increases in greenhouse gases associated with climate change. Uncertainties remain in our understanding of climate science. Nevertheless, following the July 2005 G8 summit in Gleneagles, a new consensus is emerging among major industrialised countries and even among many developing countries that, in the words of the summit's final Communiqué, "climate change is a serious and long-term challenge" and "we know enough to act now".

While there is still disagreement on how to tackle climate change between industrialised countries and developing countries, there is a common understanding that significant opportunities exist in implementing cost-effective policies to conserve energy and other resources, improve the efficiency with which we consume them and develop new carbon-saving technologies.

This joint study by the Centre for European Policy Studies (CEPS) and the Energy research Centre of the Netherlands (ECN) makes a timely contribution to the quest by governments and their supporting agencies to develop cost-effective climate change policies. The results are all the more impressive given that both institutes attempted to involve a broad constituency in the work. Preliminary findings of the study were discussed in a special workshop by a broad group of stakeholders, including academic experts, business executives, officials from the EU institutions and representatives from environmental NGOs. As Chairman of this workshop, I found these discussions not only stimulating but they also reflected the commitment of all stakeholders to grasp the methodological and data challenges implied by the concept of cost-effectiveness.

From our perspective, and given the importance that governments and the International Energy Agency attach to energy security and climate change policy, it is particularly interesting that this study has attempted to monetise the externalities, such as energy security of supply benefits and the positive effects from technology learning. In the past, governments tended to disregard long-term social costs and benefits which did not lend themselves to easy quantification.

I agree with the study that the use of cost-effectiveness criteria should be welcomed, but their application is not as straight-forward as is sometimes suggested. We at the International Energy Agency know only too well that there are widely divergent, mutually inconsistent practices in cost-benefit analysis, data issues and large cost uncertainties.

Despite or because of its careful and measured approach to cost-benefit analysis, this study will be helpful in influencing the longer-term thinking of policy-makers and policy-shapers alike. This should be well-received in light of the tremendous energy and climate change challenges that governments and the world at large face.

To make the book palatable for both policy-makers and academics, it has been divided into two parts. Part I consists of the Policy Conclusions that are drawn from the Technical Report, which presents the full analysis in Part II of the book.

Richard A. Bradley
Head of the Energy Efficiency and Environment Division
International Energy Agency (IEA)

EXECUTIVE SUMMARY

There is a growing consensus that climate change is a serious and long-term challenge with potentially irreversible consequences. The world has agreed in the United Nation Framework Convention on Climate Change (UNFCCC) to stabilise greenhouse gas (GHG) concentrations in the atmosphere at a level that prevents dangerous climate change. Given the scale of the challenge, i.e. reductions of greenhouse gas emissions in industrialised countries by 80 or 90% from today's level by the end of the century, carefully designed policies are in order that attempt to identify the most cost-effective approaches from a societal perspective.

Current designs of both national and international climate change policies today, however, tend to rest on a narrow application of social cost-benefit analysis with an emphasis on short-term efficiency of resource allocation. In contrast, this exploratory study sets out to integrate, from a societal perspective, long-term impacts of climate policy measures in the cost-benefit analysis. This is done on the basis of a literature review, combined with some own calculations. The numerical application of the proposed analytical framework focuses on ten technical measures in three different sectors: energy and industry, transport and buildings.[1]

The book is organised as follows. This Executive Summary is followed by two distinct parts. Part I summarises the policy conclusions that arise from the technical analysis. The technical analysis, including calculations, is presented in the Technical Report in Part II. Both parts are written in such a way that they can be used as stand-alone documents.

I. Policy Conclusions

The study has asked whether EU policies, such as those on security of supply, energy efficiency or support to new energy technologies, sufficiently take into account benefits in a dynamic perspective or whether these poli-

---

[1] A draft of the study was peer-reviewed (by Professor Thomas Sterner of the University of Göteborg and Dr. Felix Matthes of Öko-Institute, Berlin). The draft report was also discussed during a one-day meeting with some 30 stakeholders. Comments by the peer-reviewers and the stakeholders have been incorporated.

cies suffer from an excessive focus on short-term costs as part of the new focus on the Lisbon agenda to regain competitiveness. Typically, most climate and ancillary benefits of GHG reduction activities can only be reaped after relatively long periods, whereas the lion's share of the aggregate social costs accrues in the short term. This inter-temporal asymmetry is a key characteristic of climate change policy or more generally, sustainable development. The report also enquires whether the integration of long-term climate change objectives with other EU policy objectives such as competitiveness, security of supply, environment other than climate change, or technology policy does not fundamentally change the cost-benefit ratio and thereby moves certain hitherto non-cost-effective policies into the cost-effective camp. Some of the main results relevant for policy-making are summarised below.

1. The study proposes possibilities for the EU to use the possible presence of 'no-regrets' abatement options, which exist at least for some measures from a social, if not a commercial, point of view, in future climate change negotiations. The existence of such abatement options makes combating climate change a winning policy for most, if not all parties involved, including industrialised and developing countries. The cost-benefit analysis suggests that the level of household expenditure for energy efficiency is lower than justified by net private and social benefits; high energy prices and security of supply increases desirability of energy efficiency. The study argues that in the light of long-term climate change objectives, there is a case for stronger and more centralised energy efficiency policies. Finally, the study strongly endorses technology support to new promising energy technologies that will need to reach a critical mass in order for the production cost to fall. Such support has the potential to greatly reduce the cost of future GHG mitigation

2. From the perspective of a social cost-benefit analysis, *five* options stand out as offering the best cost-benefit ratio when taking externalities into account, at least those that can be quantified:

    •   Insulation is highly cost-effective from the end-user point of view in reducing the emissions of GHGs and has some ancillary benefits for energy security and air quality, although the overall scale for achieving reductions is only medium if compared to supply side options.

    •   Integrated gasification combined cycle (IGCC) power plants have medium costs but contribute significantly to the (probable) long-term goal of applying carbon capture and storage (CCS) in such and other coal-fired plants.

- Bio-fuels for transportation have medium to high implementation costs and high benefits for energy security, but there may be scale limitations.

- The cost of combined heat and power (CHP) is probably low (with high uncertainty), while having both a large potential to reduce emissions of GHGs, and medium ancillary energy supply security and air pollution benefits.

- Nuclear power appears to be cost-effective and has significant benefits regarding avoided air pollution and energy supply security. Yet its suitability needs to be assessed based on political acceptability and proliferation risks, and including all the costs, such as the cost of the final storage of used fuel and the risk of accidents.

The following figure attempts to graphically present the results from the calculations that were undertaken in the Technical Report. It can be found in chapter 6.2.

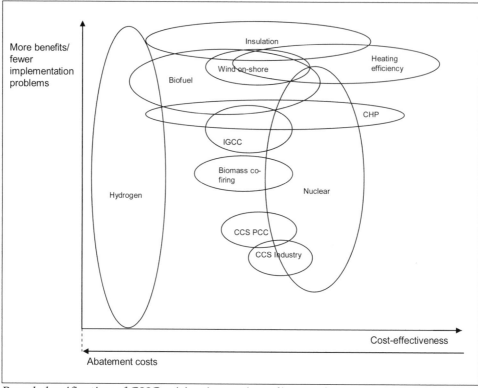

Broad classification of GHG mitigation options discussed

Notes:

- The big spread in costs for insulation results from the difference in cost approaches used, i.e. from an end user point of view or from a social perspective. In addition, the abatement costs are highly dependent on the country of implementation: in general the costs tend to be higher in Nordic countries, as many insulation measures have already been undertaken compared to other EU countries where insulation has been lacking.
- In the case of CHP, the cost range mainly arises from the sensitivity of the abatement costs to the gas price assumptions.

3. The study also indicates – implicitly – that rising energy prices do not necessarily imply that consumption falls when prices increase. The exact level of this demand reduction depends on the evolving price and income elasticities for energy services. In turn, these depend on many factors, such as sunk investment in energy-intensive equipment, the shape of the housing stock, etc. In an extreme case, a price change does not affect energy demand at all. There is evidence from

the transition countries that an increase in energy prices does not always lead to energy savings, unless consumers are put in the position to undertake energy efficiency investments and the split incentives issue.[2] Another example is the 'fuel poverty' issue, particularly but not exclusively in the UK. While a considerable part of the UK housing stock is decrepit, at the same time, simple remedial insulation measures are beyond the means of the average fuel-poverty victim or often even beyond that of the landlord's, given the time period required for any payback on the investment in terms of reduced energy bills. The past engagement of the power supplier Electricité de France (EdF) in insulation of domestic houses was motivated by the fact that it was necessary to bring domestic consumption for heat down to the point where consumers were able to pay for the total heat bill. Increased energy efficiency, which brings down the overall energy bill, becomes critical as a possible compensation tool for higher-unit energy costs. Subsidies for low-income households for energy-efficiency measures in houses can be a means of combating social exclusion through the potential lack of access to energy. This would appear to be more efficient than subsidising the consumption of low-income households.

II. Methodology Issues

The study has included externalities such as favourable impacts on air-pollution problems and energy-supply security risks in the proposed analytical framework. Quantification of externalities of air pollution is based on literature approaches, whereas a novel 'risk premium' approach is suggested for measuring impacts on the energy security of supply risk. Other externalities such as damages and employment are discussed, but their effects are however not included in the calculations because of the difficulty in quantifying them in a meaningful way.

4. To date, the (short-term) cost-effectiveness of greenhouse gas reduction options, i.e. €/tCO₂ avoided, without due regard for long-term social costs and benefits, appears to be the single-most important criterion for policy-makers in designing GHG reduction programmes. This study has shown that the application of this criterion to the job

---

[2] Insulation efforts by tenants may increase the value of the housing property concerned.

of prioritising climate change mitigation options is problematic due to:

- widely diverging, partly mutually inconsistent practices in cost-benefit analysis (CBA), the paucity of data and large cost uncertainties; and

- its disregard for many long-term social costs and benefits in which quantification problems constitute but one (important) underlying factor.

5. The results that the study have given in monetary value show that taking into account ancillary externalities can significantly change the net social cost of certain mitigation options and therefore the net social cost of climate change policies. In addition to externalities, key factors having a high-cost impact are the discount rate(s) used and energy price trajectories over time. These and other cost uncertainties must be duly taken into account in preparing cost-effectiveness analysis of climate change mitigation options and policy-making.

6. As to more general climate policy design, the following methodological conclusions should be kept in mind:

- It is important that interactions of different policy options are reviewed to make sure that options retained for policy implementation purposes are not incompatible with each other.

- Efficiency prices (i.e. by and large, market prices net of taxes and subsidies) should be used as a point of departure for cost-benefit analysis from a societal point of view.

- The analysis should apply explicitly the context-specific suitable discount rate without 'automatically' applying discount rates used by authoritative economic development analysis and planning bodies.

- Uncertainties surrounding resulting key figures regarding mitigation costs per option should be shown quantitatively.

- Serious efforts to quantitatively include major external costs and benefits in resulting key figures are needed.

PART I
POLICY CONCLUSIONS

C. EGENHOFER

AND

J. JUSSILA HAMMES

POLICY CONCLUSIONS

1. Introduction

For several years now, the EU as well as many other developed and developing countries have identified climate change as among one of the world's most important challenges, and have accordingly been engaged in developing cost-effective policies aimed at climate change combating. There is a growing consensus among all major industrialised countries and even among many developing countries that "climate change is a serious and long-term challenge" related strongly to the increased use of "fossil fuels and other human activities".[1] That same communiqué concluded: "We know enough to act now".

From the beginning, a central element of climate change policy has been cost-effectiveness. Cost-effectiveness concerns have been one of the principal drivers behind the Kyoto Protocol's flexible mechanisms as well as the EU and other emissions trading schemes. Cost-effectiveness has been the 'leitmotiv' of the European Commission's efforts in following through the formulation of EU policies within the European Climate Change Programme (ECCP), the EU emissions trading scheme (ETS), the European Commission's 2005 Communication and the review of the European Climate Change Programme to be completed in 2006. Finally, in 2005, the European Council asked the European Commission to continue its work on assessing the costs and the benefits of medium- to long-term climate strategies.

It is in this context that CEPS together with the Energy research Centre of the Netherlands (ECN) has undertaken an extensive literature review, combined with our own calculations, on the social costs and benefits of climate change mitigation options with the aim of informing crucial EU

---

[1] Citation from Gleneagles communiqué of the G8 summit in July 2005, opening section.

policy processes.[2] On the costs side, the accompanying Technical Report contributes to the existing literature by taking into account new aspects, such as the need to adjust the discount rates according to the type of analysis conducted, energy price trajectories, the need to include various ancillary costs in the analysis and the EU's long-term climate-change aspirations. On the benefits side, the EU has attempted to monetise a number of externalities, including environmental benefits other than GHG mitigation, energy security of supply benefits and the positive effects from technology learning. Furthermore, the report discusses the inclusion of a number of other factors that affect the costs and the benefits of the greenhouse gas policy, notably avoided damages and employment. The latter effects are, however, not included in the calculations because of the difficulties in quantifying them in a meaningful way. Most robust benefit estimates were obtained on the environmental and – possibly to a lesser extent – on energy security questions. Other externalities were not monetised. The details of the findings are presented in the Technical Report that follows in Part II.

The study has asked whether current as well as soon-to-be formulated EU policies, such as those on security of supply, energy efficiency or support to new energy technologies, sufficiently take into account benefits in a dynamic perspective or whether policies suffer from an excessive focus on short-term costs as part of the new focus on the Lisbon agenda to regain competitiveness. It also wonders whether the integration of long-term climate change objectives with other EU policy objectives, such as competitiveness, security of supply, environment other than climate change, or technology policy, does not fundamentally change the cost-benefit ratio and thereby move certain, hitherto non-cost-effective policies into the cost-effective camp. In particular, the question can be raised whether in the light of long-term climate change objectives, there is not a case for stronger and more centralised energy-efficiency policies. The study further proposes possibilities for the EU to use the possible presence of 'no-regrets' abatement options, which exist for some measures – if not from a financial point of view then at least from a social one – in future climate change negotiations. The existence of such abatement options would make combating cli-

---

[2] For instance, the European Climate Change Programme (ECCP), the continued work by the European Commission on the costs and the benefits as mandated by the European Council or work related to the Green Papers on energy efficiency and energy policy.

mate change a winning policy for society at large even when the climate change issue is disregarded in industrialised and developing countries alike. Finally, the study also makes the case for technology support to new promising energy technologies that will need to reach a critical mass in order for the production cost to fall. Such support has the potential to greatly reduce the cost of future GHG mitigation.

2. Is the EU losing sight of the benefits of environmental policy?

In looking at the history of the EU's environment and climate policy, it is possible to discern two broad phases. There was the early, innovatory or 'evangelist' phase roughly spanning the mid-1980s to the mid-1990s. This was followed by an implementation or 'performance' phase where the cost of environmental policies has become more important (e.g. European Commission 2006a).

During the first phase, environment and climate policy were seen as an opportunity for beneficial change to improve the efficiency of the economy. Similar thinking on the environment and climate change prevailed in other OECD countries (for a summary overview see Fujiwara et al., 2006; see also OECD, 1996 & 1997). Central to this tradition is the idea that many current economic activities over-produce economic 'bads' or external costs in the form of costs imposed by one private economic actor on others without regard to the latter's well-being, for instance air and water pollution. Such externalities exist because the firms and individuals giving rise to them face only the private cost of their actions, not the full social cost incorporating the full cost of the polluting activity. This approach embraces the notion that environmental policies, broadly speaking, enhance welfare and are therefore beneficial. This approach to environmental policy has further been associated with the concept of environmental tax reform. At the highest political level, this notion was put forward by the 1993 Delors White paper on Competitiveness, Growth and Employment (European Commission, 1993).

The second phase of the response has increasingly seen environmental initiatives, such as in the field of climate change or in chemicals regulation, as a threat both to the economic prospects of European nations, and to the status quo. In such a perspective, energy or carbon taxes, liability rules or tradable permit schemes are presented as an economic distortion – which is invariably costly. This approach tends to focus on the cost of adjustment (see European Commission, 2006a). While the initial Lisbon agenda has identified environmental protection as a source of economic growth, with the rise of the real, or perceived, 'loss of competitiveness' of the EU, the importance of short-term costs have increased.

Much of the EU's climate change policy *formulation* has adopted the first approach as evidenced by the successive phases of the European Climate Change Programme (European Commission, 2001; ECCP I review working group under ECCP II).[3] This may have initially been encouraged by a desire to use tax revenues to reduce employment taxes, helped by the relatively mild Kyoto Protocol targets (compared e.g. with those of Japan, Canada or the US). This desire was complemented by a generally strong preference for multilateral approaches, which helped to unite the EU at the Gothenburg European Council in the light of the brusque rejection of the Kyoto Protocol by the US President Bush. Consequently, the approach helped to generate political momentum. This relative consensus was also able to achieve the implementation of the EU CO_2 emissions trading scheme, which has enjoyed the broad support of industrial stakeholders along with the governments and environmental non-governmental organisations (NGOs).

Nevertheless, with progressive implementation of the EU's climate change policies, the first approach has increasingly come under pressure, mainly by industrial interest groups and by some governments subject to interest group pressure. This new focus is probably in part the result of the EU's current relative isolation when it comes to implementing constraining climate change policies and of the distributional imbalances that are felt within the EU emissions trading scheme (see e.g. IEA, 2005; Carbon Trust, 2004). But climate change is not the only area where environmental legislation has been under attack. The same has been true for the European Commission's proposal to streamline the registration and authorisation procedures for chemical substances (REACH) and for the Commission's proposals on air quality regulation.

This raises the question of whether the EU, which made much initial progress in setting up the international momentum for climate change mitigation in the first (evangelist) phase, in the second (implementation) phase, has lost confidence in the strong 'benefits' position. It is further possible that the EU, during the era of low energy prices backed away from correcting market failures – in particular as we have seen in the areas of

---

[3] For the reports of the European Climate Change Programme, see http://ec.europa.eu/environment/climat/eccp_review.htm.

transport and domestic energy use – and now, in another era of much higher energy prices is unable to fulfil its commitments. This perceived lack of confidence in correcting market failures appears all the more surprising as market as well as political conditions such as worries over energy security and climate change are now helpfully pushing energy policy in the direction of internalising externalities.

3.   The daunting task of meeting the climate change challenge

The Intergovernmental Panel on Climate Change (IPCC) in its Third Assessment Report (IPCC, 2001) warns that an increase in global temperatures is likely to trigger serious consequences for humanity and other life forms, including a rise in sea levels, which will endanger coastal areas and small islands, and a greater frequency and severity of extreme weather events (e.g. Schellnhuber et al., 2006; EEA, 2004).

As early as in 1996, the EU adopted a long-term target of limiting the temperature increase to a maximum of 2°C.[4] In order to have a reasonable chance of achieving this, the CO_2 concentration levels would need to stabilise below 550 ppmv CO_2 equivalent or 450/475 ppmv CO_2 only.[5] This would likely require a peak of global emissions before 2020 (IPCC, 2001), since GHG emissions stay in the atmosphere for a long time (see Table 1),[6] making a strong case for taking action now. Waiting longer will make

---

[4] See Conclusions of the Council of the European Union, meeting in Luxembourg in June 1996 (Council of the European Union, 1996). This was recently reiterated by the Environment Council in December 2004 with reference to the IPCC's Third Assessment Report and reaffirmed by the European Council in March 2005: "the overall global mean surface temperature increase should not exceed 2°C above pre-industrial levels" (European Council, 2005). It is however uncertain whether this target will be sufficient to actually avoid 'serious consequences', as climate sensitivities are high and there is still much we do not know about climate change.

[5] In comparison, pre-industrial CO_2 concentration levels stood at 280 ppm, while they have increased to 377 ppm to date, leading to an increase in the average global temperature by almost 1°C (Council of the European Union, 2004). In the absence of measures, there will be no stabilisation below 700 or even 1,000 ppm. Such levels, according to the IPCC, are likely to lead to very damaging impacts, including structural alterations to weather patterns or even to changes in important ocean currents, such as the Gulf Stream.

[6] For example, CO_2 stays in the atmosphere for more than 100 years. What we emit today will cause damage for a long time in the future.

reaching the target more difficult and/or reaching certain stabilisation trajectories impossible.

Table 1. Conditions for stabilisation of CO_2

WRE CO_2 stabilisation profiles (ppmv)	Accumulated CO_2 emissions 2001 to 2100 (GtC)	Year in which global emissions peak	Year in which global emissions fall below 1990 level
450	365-735	2005-2015	<2000-2040
550	590-1135	2020-2030	2030-2100
650	735-1370	2030-2045	2055-2145
750	820-1500	2040-2060	2080-2180
1000	905-1620	2065-2090	2135-2270

Source: IPCC 2001 – Third Assessment Report – Synthesis Report, 2001.

The task of achieving long-term climate change objectives is daunting. World energy demand is projected to grow by around 60% or even more by 2030 (IEA, 2004; European Commission, 2003; ExxonMobil, 2004). Until 2050, global energy demand will double or possibly even triple (WBCSD, 2004). The main drivers of increasing global energy demand are economic development and projected population growth in developing countries.[7] It is further realistic to assume that the EU and the world at large will continue to rely on fossil fuel as the principal fuels for the time being.[8]

For example the World Business Council for Sustainable Development, a coalition of 180 multinational companies with focus on sustainable development, using IPCC scenarios, assumes that in order to achieve stabilisation of GHG concentrations at a level of 550 ppm CO_2,[9] there is a need to reduce global CO_2 emissions by around 6-7 bn tonnes (gigatonnes) of carbon (or 22-25 billion tonnes of CO_2) per year by 2050 as compared to a

---

[7] World economic growth is expected to average around 3% annually while the population will grow at an average of 1% per year according to most forecasts. Hence population could increase to 9 billion by 2050 (see UN, 2004).

[8] The IEA (2004) assumes that fossil fuels will continue to dominate global energy use in 2030, accounting for some 85% of the *increase* in world energy demand. Total global CO_2 emissions are expected to grow to over 50 billion tonnes.

[9] The EU assumes a 450 ppm – hence more stringent – stabilisation target.

situation where no policies are put into place.[10] Such a no-policies scenario assumes that total global emissions would increase from current 9 billion tonnes of carbon (33 billion tonnes of CO_2) to more than 14 billion tonnes of carbon (51 billion tons of CO_2) This reduction would equal around 70-80% of current total global emissions (see Figure 1 and WBCSD, 2004). In comparison, the overall EU-15 target in the Kyoto Protocol has been around 111 million tonnes of carbon (or 400 million tonnes of CO_2).[11]

Figure 1. Achieving an acceptable CO2 stabilisation

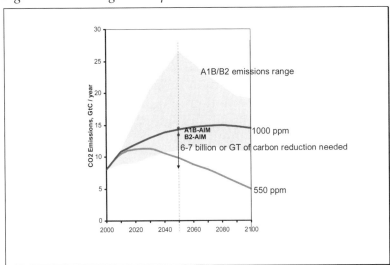

Notes: 1 GT = 1 billion tonnes.

6-7 Gt of carbon is equal to 22-25 billion tonnes of CO_2.

A1B-AIM/B2-AIM are IPCC scenarios used by WBCSD; B2 describes the lower energy-use scenario, i.e. intermediate level of global growth while A1B describes the higher energy use scenario, i.e. very rapid global economic growth.

Source: WBCSD (2004), based on scenarios from the IPCC's Third Assessment Report (IPCC, 2001).

---

[10] One tonne of carbon equals 3.67 tonnes of CO_2.

[11] The current gap to meeting that target is around 200-300 million tonnes of CO_2, according to EEA (2005), EEA Report 8/05. The figures change depending on the reference scenario.

To illustrate the scale of the task, any one of the following activities could be implemented in order to achieve reductions of 1 billion tonnes of carbon emissions (or 3.67 billion tonnes of CO_2): an increase of 150 times the current wind power capacity, the installation of five times the current nuclear capacity or 1 billion hydrogen cars could be brought into operation to replace conventional cars obtaining fuel economy of 8 litres per 100 kms. Alternatively, one could use half of the entire US agricultural area for biomass production. To cite measures from the building area, this would mean approximately 9 times the savings from the hypothetical global application of the EU Directive on Energy Performance in Buildings. Note that the above illustrations are very tentative and should not be taken as exact numerical examples.

Table 2. Reducing 1 billion tonnes of carbon (= 3.67 billion tonnes CO_2) per year requires…

Technology	Required for 1 bn T reduction of carbon
Coal-fired power plant with CO_2 capture & storage	700 x 1 GW plants
Nuclear power plants replace average plant	1500 x 1 GW (5 x current)
Wind power replaces average plant	150 x current
Solar PV displace average plant	5 x 1 million (2000x current)
Hydrogen fuel	1 billion H_2 cars (CO_2-free H_2) displacing 1 billion conventional 30 mpg (approx. 8 litres per 100 kms) cars
Geological storage of CO_2	Inject 100 mb/d fluid at reservoir conditions
Biomass fuels from plantations	100 x 1 million ha (half of US agricultural area)
Measures related to energy consumption in existing building stock (heat installation, insulation, appliances, etc.)	9 times the EPB Directive for EU-15 if extended to all houses (hypothetical global application)

Notes: The above illustrations are based on an assumed stabilisation at 550 ppm (The EU assumes a 450 ppm – hence more stringent – stabilisation target).

'Average plant' describes current fuel mix.

Mb/d = million barrels per day.

Source: Egenhofer & van Schaik (2005), p. 8 (updated).

While there are different opinions on whether medium-term climate change goals can be reached with conventional technology,[12] in the long-term (i.e. beyond 2050), the development and diffusion on new and technically unproven (i.e. breakthrough) technologies will be crucial to meet the UNFCCC's objective of stabilisation.

4. EU energy and climate change policies

Achieving long-term climate change objectives tends to rely on action within five principal areas: direct combustion in industry, power generation, mobility, consumer choices/lifestyles and buildings. Advances in emissions reductions are attained by a combination of improved efficiency in energy production and consumption based on incrementally improved technologies and techniques and the development and diffusion of new and yet unproven technologies. Over time, the gains by developing and deploying new technologies overtake efficiency gains by improving existing technologies and techniques as the principal source for emissions reduction.

4.1 Energy efficiency

As new technologies take time to develop, reductions in the short-term are likely to be achieved by increasing the efficiency of existing technology solutions and by accelerating technology diffusion, i.e. to encourage the use of the most efficient technologies. In the World Alternative Policy Scenario formulated by the IEA in its 2004 World Energy Outlook, more efficient use of energy in a wide range of applications, including vehicles, electric appliances, lighting and industrial uses, accounts for almost 60% of the reductions in CO_2 emissions.[13]

The energy efficiency potential has been addressed in a number of EU strategic initiatives. These include the European Commission's climate change Communication from February 2005 and the European Council Presidency Conclusions from March 2005. The Green Paper on energy effi-

---

[12] Pacala & Socolow (2004) and IPCC (2001) argue that the climate problem could be solved for the next 50 years with current technologies, whereas Hoffert et al. (2002) maintain that new and revolutionary technologies would be needed.

[13] A shift in the fuel mix for power generation in favour of renewables and nuclear energy power accounts for most of the rest (IEA, 2004).

ciency by the European Commission (European Commission, 2005c) led to the formulation of the Energy Efficiency Action Plan in October 2006 (European Commission, 2006d). Energy efficiency has also been a central element of the February 2006 Green Paper on energy policy (European Commission 2006b, 2006c) and the March 2006 European Council Presidency Conclusions on Energy Policy for Europe.

The principal initiative of the EU has been the 2005 Green Paper on energy efficiency in combination with the Energy Efficiency Action Plan, however. The analysis and prescription focused to a large extent on realising 'cost-effective' savings to boost efficiency and competitiveness in Europe as a whole and for the European industry in particular. The Green Paper's main targets are the many barriers to energy efficiency that have been identified in the literature (e.g. IPCC, 2001).[14] A clear emphasis has been on the Lisbon agenda of making Europe more competitive. While there are many references to short- and long-term climate change and security of supply issues, this strong emphasis on competitiveness begs the question whether climate change benefits have been underestimated. The analysis in the Technical Report suggests that this is the case.

4.2   Technology development and diffusion

The long-term climate change perspective has been far stronger in the March 2006 Green Paper on energy policy (European Commission, 2006b) than in for example the energy efficiency Green Paper. In fact, the former establishes for the first time the irrevocable link between security of supply and climate change. The overarching theme of the energy Green Paper is how to ensure 'secure' and 'low-carbon' energy supplies for the future.

The analysis in the Technical Report in Part II points to the importance of adapting a dynamic technology development approach, where the cost of future technology is not fixed but depends on other interacting technology developments and above all, on the policies adopted today. It argues that "from a sustainable development and long-term energy supply security perspective, a very high priority is warranted to put in place proper policy frameworks that foster acceleration of exhaustible-resource-

---

[14] The most significant barriers are a fragmented market structure (small firms, different types of buildings, many stakeholders), split incentives between owners and renters, capital constraints, information gaps/asymmetry and complexity, and finally the poor availability of climate-friendly appliances.

saving innovations." The climate change issue has only enhanced the urgency for human kind to accelerate sustainability-enhancing technological development.

It is particularly important to acknowledge interaction between different technology options. Some materials or processes may have higher GHG emissions or other negative environmental effects in the production process, which however are overcompensated for by the higher (environmental) efficiency of the final product. Examples are low-sulphur motor fuels that have high GHG emissions at production, light-weight steel and aluminium that reduce car emissions by making cars lighter and therefore more fuel-efficient, or the use of certain GHG gases in order to improve insulation in double-glassing.

Technologies can also be mutually dependent. For example the development of carbon capture and storage technologies may depend on the progressive deployment of IGCC power plants. Therefore, technological choices often affect development of other technological options.

A third important aspect of technology development is learning-by-doing. The analysis in the Technical Report shows that implementing technologies 10 years later may change the marginal production costs, and consequently, the technologies' cost-effectiveness significantly. In the Technical Report this effect is most notable for wind power and for IGCC. However, the learning-by-doing calculations are only valid provided the expected learning rates are achieved. The projected learning rates are based on an extrapolation of past learning rates for other energy technologies in the past. Nevertheless, the analysis points to large possibilities to future cost savings if the energy production technologies are supported by official means in their earlier years.

This means that subsidies for certain – in a climate change perspective – relevant technologies need to be judged in a dynamic rather than static perspective. The question should not only be what does it cost today but also what possible benefits can we derive tomorrow. One caveat needs to be made however; while this basic reasoning is robust, the nature of the problem only allows for a quantitative assessment when rather sweeping assumptions are made on key parameters such as the learning rate.

One caveat needs to made, however, which is the focus of a long-running dispute with energy economists and has profound implications for energy and climate policy; the rebound effect. The rebound effect generally distinguishes three variants: direct, indirect and economy-wide rebound effects.[15] Direct rebounds describe the possibility that improved efficiency should reduce the price of supplying an energy service, which in turn should increase consumption of that service. Entirely analogous effects are applicable to improvements in energy efficiency by manufacturers (economy-wide rebounds). Furthermore, a fall in the real price of energy services will reduce the price of products throughout the economy and lead to a series of adjustments, with energy-intensive goods and sectors gaining at the expense of less energy-intensive ones. Energy-efficiency improvements should also increase economic growth, which should itself increase energy consumption by some second-order fraction.

5. Implications for policy-making

Within the EU and elsewhere, cost-effectiveness of greenhouse gas reduction options, i.e. ensuring that the direct cost in euro per tonne of CO_2 ($€/tCO_2$) avoided is the lowest possible, appears to be virtually the single most important decision criterion for policy-makers in designing GHG reduction policies. While this is to be welcomed in principle, the application of the cost-effectiveness criteria is not as straightforward as it is sometimes suggested. There are widely diverging, partly mutually inconsistent cost-benefit analysis (CBA) practices, paucity of data and large cost uncertainties. More importantly, as the Technical Report points out, many long-term social costs and benefits tend to be disregarded in which quantification problems constitute but one (important) underlying factor. The principal quantification issues include the following:

- the perspective from which cost and benefits are (tacitly or explicitly) valuated,
- the time horizon considered,
- the rate at which costs and benefits are discounted,

---

[15] An example of direct rebounds is that a more efficient heating system may allow higher levels of thermal comfort to be enjoyed. This increase in consumption will partly offset the energy savings that are achieved. An indirect rebound is for example, that the savings from lower heating bills may be put towards a far-away holiday. Such additional spending will involve the consumption of energy, and this will further offset the energy savings achieved.

- the extent to which non-climate ancillary costs and benefits are included in the analysis and

- uncertainties surrounding the various costs and benefits.

The Technical Report points out that in addition to externalities, the applied discount rate(s), energy price trajectories over time and the uncertainty about the future cost of the energy production technology are the principal determining factors for cost estimates. Hence, they exhibit a high sensitivity to variations in these factors. Cost uncertainties should be duly taken into account in preparing cost-effectiveness analysis of climate change mitigation options and policy-making.

5.1 Priority options

There is a consensus within the EU and the European Commission that post-2012 climate policy should aim broadly at keeping mitigation costs low and at strengthening the cost-effectiveness of climate policies as one of the main pillars of a European Climate Change Programme. This was one of the reasons that the European Council in March 2005 asked the European Commission to continue its work on the costs and the benefits of medium- and long-term climate strategies. Unfortunately the European Council did not indicate which particular strategies and targets should be the subject of this analysis (European Commission, 2005b). Among the options considered, the Technical Paper has identified five options that are set to show robust net benefits when taking into account externalities.

- Insulation is very cost-effective from the end-user point of view in reducing GHG emissions and has some ancillary benefits for energy security and air quality, although the overall scope for achieving reductions is only medium if compared to supply options.

- Integrated gasification combined cycle (IGCC) power plants have medium costs but contribute significantly to the (probable) long-term goal of applying carbon capture and storage (CCS) in such and other coal-fired plants.

- Bio-fuels for transport have medium-to-high implementation costs and high benefits for energy security; there may be scale limitations.

- The cost of combined heat and power (CHP) is low while having both a large potential to reduce GHG emissions, and medium ancillary benefits on energy supply security and air pollution.

- Nuclear power appears to be cost-effective and has significant benefits regarding avoided air pollution and energy supply security. Yet its suitability needs to be assessed based on political acceptability and proliferation risks, and all costs need to be included, such as the cost of the final storage of used fuel and the risk of accidents.

5.2 Energy efficiency is where global cooperation is possible

The EU's short-run emissions target is given by the Kyoto Protocol. In the long run, the objective is, for example, spelled out in the 2005 Spring European Council Conclusions. Both in the short- and in the long-run, the principal policy tools to meet the objectives depend on an increase in the efficiency of both the demand and supply of energy. As noted in Section 4.1, energy efficiency, especially in the short run, can be achieved by incremental improvements (i.e. competition-driven innovation that promise rents) and by the accelerated diffusion of existing technologies.

Also from a global perspective, energy efficiency is the most promising area where cooperation on climate change can be achieved. The Gleneagles Plan of Action on climate change, clean energy and sustainable development to be implemented under the auspices of the World Bank and the International Energy Agency (IEA) has identified energy demand as the principal area for immediate action. There is an explicit reference to co-benefits including greenhouse gas emissions but also local and regional pollution, health, security of supply, competitiveness and employment benefits – areas that have also been analysed in the Technical Report.

In the Gleneagles Plan of Action, buildings are singled out as the first area for energy-efficiency improvements. The focus is the review (by the IEA) of existing building standards and codes, the development of energy indicators and the identification of best practices (p. 1). The other priority areas for action are appliances, surface transport, aviation and industry. Within the US portfolio of action on climate change, energy efficiency in the form of standards or other measures plays the principal role in the short-term.[16]

While this points to a consensus on the crucial importance of energy efficiency, the results of the Technical Report show that more could be done and in fact would make sense in the light of long-term targets. The Techni-

---

[16] E.g. fuel economy standards, energy efficiency standards as well as other regulatory programmes or tax breaks for energy supply, industry or land use; see White House press release, "President Bush is addressing Climate Change", 30 June 2005.

cal Report provides further economic arguments for energy efficiency policy; considerable no-regret options may exist and may be understated, included among them insulation. The Technical Report confirms that insulation, when compared to other options, is very cost-effective from the end-user point of view and has certain benefits for energy security and air quality. It suggests that the level of household expenditure is lower than justified by benefits, even accounting for uncertainties regarding costs.

Benefits from energy efficiency are particularly associated with energy prices (currently high) and with security of supply. The higher the prices and, more importantly, the greater fears about energy security are, the more desirable energy efficiency becomes. In economic terms, this could be expressed as if a higher risk for supplies raises the discount rate. The discount rate rises since uncertainty increases the required social risk premium, which is added to the risk-free discount rate used to discount energy consumption. A higher discount rate has as a consequence a lower present value of future energy consumption, which should have a dampening effect on consumption, which, in turn enhances energy security of supply. From a social point of view, investments that improve energy efficiency whether by improving the insulation, the purchase of new machines or by other means has the desired aggregate effect not only of reducing GHG emissions but also of providing additional security of supply.

5.3 Energy efficiency can partly finance itself but may also need to be subsidised

In addition to social benefits, such as GHG and air pollution mitigation and benefits to the security of supply, investments that improve energy efficiency, whether by improving the insulation or by making the choice of efficient appliances or fuels more enticing, tend to yield benefits in the form of lower energy bills. This is particularly important as we should expect that new energy technologies to combat climate change or policy measures such as the EU emissions trading scheme or national taxes will increase both the wholesale and the retail energy prices. The Technical Report indicates that the damage cost of climate change emissions is likely to increase over time. The Technical Report assumes that the marginal cost of GHG reduction can well exceed 100 €/tCO_2-eq. Since substantial ancillary benefits of GHG reduction options often exist, benefits from air pollution reduction and security of supply can offset a large part of the financial costs from a social perspective. Nevertheless, from an end-users' point of view, the cost for energy is likely to go up as a result of climate change policy.

This would normally mean that consumption falls when prices increase. The exact level of this demand reduction depends on the price elasticity of energy demand, which depends on many factors, such as sunk investment in energy-intensive equipment, the shape of the housing stock, etc.[17] In an extreme case, the demand function may, however, be vertical (which indicates totally inelastic demand), so that a price change does not affect energy demand at all. This is for instance the case found by the fuel poverty literature, particularly but not exclusively in the UK. For example, a considerable part of the UK's housing stock is decrepit. At the same time, simple remedial insulation measures are beyond the means of the average fuel poverty victim or often even of the landlord's, given the time period required for any payback on the investment in terms of reduced energy bills. There is further evidence from the then-transition economies of the former Soviet bloc, which demonstrates that increases in energy prices do not automatically lead to demand responses (e.g. EBRD, 2001). Consumers need therefore to be put into the position to undertake energy efficiency investments. Otherwise the result will be non-payment, especially if the overall energy bill rises beyond around 15% of total disposable household income (EBRD, 2001). While energy-education schemes could help, transition economies typically lack the institutional structure to make good use of them. The EBRD report concludes that while price tariff reform is essential for efficiency gains on the consumption side, it must be accomplished alongside the establishment of appropriate help for vulnerable households. Similarly, and as a further example, in the past the electricity supplier Electricité de France (EdF) was actively engaged in insulation of domestic houses, which was motivated by the fact that it was necessary to bring domestic consumption for heat down to the point where consumers were able to pay for the total heat bill.

This raises the question of whether energy efficiency policy, i.e. measures including subsidies for refurbishment, are not a better tool to address the market failures than to provide transfers to vulnerable customers,

---

[17] Even the supply side plays an important role in the changing consumption of energy. According to economic theory, supply is an increasing function of price. The climate policy works by raising the production cost of energy, which is the underlying reason behind the prise rise for consumers. The (price) elasticity of energy supply is determined by several factors, such as market failures and the possibility to substitute fuels, but also in the long term on the possibility to build new generation capacity.

which has been the case in some member states after the recent price increases. Providing transfers to vulnerable consumers amounts to little more than subsidising consumption with no effect on demand. By extension, this policy has no effect on the GHG emissions or on the security of supply. In this line of thinking, subsidies for low-income households for energy-efficiency measures in houses can be a means of combating social exclusion through the potential lack of access to energy, as has for example been described by the Socialist Group in the European Parliament in their draft position paper on a sustainable common energy policy for Europe.[18]

5.4 Support to new technologies lowers the future cost of emissions reductions

Energy efficiency policy is a demand-side policy. A (cost)-efficient climate change policy attacks the problem both from the demand and from the supply side, however. One supply side measure emphasised in the Technical Report is the support to be given to the introduction of new, promising energy-production technologies. Assuming that the new production technologies present learning-by-doing effects of the same type as older technologies, it can be expected that a doubling of the generation capacity lowers the investment costs by a certain percentage. This may have a considerable impact in lowering the cost of GHG mitigation policies in the future, and although present emissions reductions may be relatively modest, they may save the process in the future.

The report especially highlights the benefits from investing in IGCC and wind power generation, presenting calculations, which are based on certain assumptions, on the cost reduction that can be reached for these generation technologies by the year 2020 if the investment in the technologies takes off in 2010 rather than not until 2020. These cost reductions can be considerable. Considering the poor track record of governments in choosing which technologies to support, mechanisms would however have to be created that helped them to choose the technologies to support. An

---

[18] p. 5 in the section on buildings of draft version dated 16 June reads: "Considering the need to tackle the social consequences of high energy prices, member states' actions should particularly support low-income families and individuals to achieve energy savings in their homes, thereby reducing their energy bills and their exposure to future price increases. This can be partly financed through the European Regional Development Fund."

alternative is to divide the available funds thinly over all new energy-production technologies. This latter alternative may however be even more counter-productive than investing heavily in technologies that turn out to be duds, since it risks too low support levels to all technologies in order to reap any real gains from the subsidies.

European cooperation in this area is warranted in order to avoid unnecessary overlapping of subsidies, and consequently, research and demonstration. Coordination is already evident in the EU's distribution of research funds, but cooperation between the member countries should be deepened. Furthermore, not only support for basic research is called for here, but also to demonstration projects and continued support in the early markets for promising technologies. Whereas the former is often undertaken by universities and research centres within the public sphere, the latter is often left to the energy-production companies. These companies are probably well aware of each other's activities and therefore, unnecessary duplication of efforts is less likely to occur. Nevertheless, if the EU and the member states support such activities, coordination of subsidies at the EU level would be warranted.

5.5   Is there a case for more energy efficiency at EU level?

The previous sections have made the case for energy efficiency policy, mainly but not only as a result of market failure. This has been the focus of the Commission's Green Paper on energy efficiency and the Energy Efficiency Action Plan (European Commission, 2006d). The Green Paper has identified options that could save up to 20% of energy consumption or €60 billion annually in a cost-effective way. While this is an impressive figure, the Technical Report has made the case that cost-effectiveness is very difficult to assess and is particularly dependent on assumptions on price trajectories, discount rates, technology learning effects and, in an extreme case, even on the external costs and benefits of actions. Hence, more action could be optimal from a social perspective, depending on the assumptions adopted. The following paragraphs will provide strong arguments that there is a case for a more centralised EU energy efficiency policy.

As there are only rudimentary competencies for energy proper in the EC Treaty, EU intervention is typically based on internal market or environment rules. For the domain of the internal market, essentially this means to ensure free movement of goods, services, persons and capital. As a result, there are EU-wide technical standards for tradable goods, typically including environmental provisions. Intervention in the field of the environment is a shared competence, falling under the subsidiarity principle that says that EU action should only be undertaken if there is an added-value for reasons of economies of scale or (positive or negative) externalities. For energy efficiency this typically has meant EU-wide measures for raising awareness, the promotion of best-practice as well as obligations for member states to undertake certain actions, without prescribing specific measures or targets.

This appears to have been justified during the period of relatively low energy prices and the absence of GHG emissions reduction targets, which has meant low importance given to energy efficiency. With relatively little importance given even to energy security, member states insisted that energy policy remained in their hands. More recently, however we have seen a change of position. Starting with the Hampton Court informal summit and the Spring 2006 European Council, there is a growing awareness that the security of energy supply has an EU dimension and that the EU should develop an energy policy for Europe. As we have seen, the two most immediate and least-contested elements from an environmental or climate change policy point of view of such a policy are demand-side measures and support to technology development. Indeed, if the EU wishes to undertake a common policy on supply, there is an even stronger argument for a common policy on demand. This would be consistent with the EU's drive towards an internal market for energy.

As this study has made clear, the unprecedented challenges that long-term climate policy will pose provide further urgency to a more forceful EU energy policy. Reductions in the order of up to 80% or even 90%, as compared to the 1990 GHG emissions levels by 2100 necessitate a consistent approach to member states' energy policies, e.g. common approaches to renewables support, carbon capture and storage, nuclear, or energy efficiency. Huge discrepancies between member states will not only risk creating distortions to competition or barriers to cross-border trade but also possibilities for beggar-thy-neighbour policies. And since it is not unreasonable to expect an EU of 27 or more member states not to create another GHG burden-sharing agreement, it will be the EU that will be responsible for reaching any climate change targets it is likely to eventually sign up to. But there are even more immediate reasons for a more harmonised if not centralised EU approach to energy efficiency. Domestically, it will be of crucial importance that the EU achieves GHG reductions across all sectors, including notably buildings and transport as well as industry and energy supply to avoid the risk that emissions reductions will have to be undertaken mainly by the sectors covered by the ETS. Excessive reliance on the ETS sectors could increase the cost burden of industry and in turn might undermine industry's competitiveness (see Egenhofer & Fujiwara, 2006). If the EU wants to continue its leadership in international negotiations, it will not only need to prove that it can reduce its emissions but also that this can be done in a cost-effective way and maintain political acceptability. The Technical Report indicates that energy efficiency in buildings is set to be prominently represented in an efficient portfolio of climate change policies and measures. It will also be greatly enhanced by an early introduction of innovative energy-production technologies. Finally, as we have pointed out on previous occasions, energy efficiency to date is one of the most promising areas for international cooperation on climate change. Such cooperation is likely to be facilitated if the EU increasingly speaks with one voice.

REFERENCES

Carbon Trust (2004), *The European Emissions Trading Scheme: Implications for Industrial Competitiveness*, Carbon Trust in the UK, June.

Council of the European Union (1996), 1939th Council Meeting, Luxembourg, 25 June.

-------- (2004), Press Release of the 2632nd Council Meeting, 20 December, 15962/04 (Presse 357).

EBRD (2001), *Transition Report 2001*. European Bank for Reconstruction and Development. London

EEA (2004), *Impacts of Europe's changing climate: An indicator-based assessment*, EEA Report No 2/2004, European Environment Agency, Copenhagen.

EEA (2005), EEA Report 8/05, European Environment Agency, Copenhagen.

Egenhofer, C. and L. van Schaik (2005), *Towards a Global Climate Regime: Priority Areas for a Coherent EU Strategy*, CEPS Task Force Report, Centre for European Policy Studies, Brussels, May.

European Commission (1993), *White Paper on growth, competitiveness, and employment: The challenges and ways forward into the 21st century*, COM(93) 700 final; Brussels, 5 December.

-------- (2001), *European Climate Change Programme*, Final Report (http://www.europa.eu.int/comm/environment/climat/eccpreport.htm).

-------- (2003), World energy, technology and climate policy outlook, WETO 2030, Community research, Office for Official Publications of the European Communities, Luxembourg.

-------- (2004), *European Competitiveness Report 2004*, Commission Staff Working Paper SEC(2004) 1397 of 8 November.

-------- (2005a), *Winning the Battle against Global Climate Change*, Communication from the Commission to the Council, the European Parliament, the European Economic and Social Committee and the Committee of the Regions, COM (2005) 35 final.

-------- (2005b), *Winning the Battle Against Global Climate Change – Background Paper*, Commission Staff Working Paper, Brussels, 9 February.

-------- (2005c), *Doing more with less*. Green Paper on energy efficiency, 22 June.

-------- (2006a), First report of the High-Level Group on competitiveness, energy and the environment, 2 June.

-------- (2006b), A European Strategy for Sustainable, Competitive and Secure Energy, Green Paper, Com(2006) 105 final, 8 March.

-------- (2006c), Annex to the Green Paper, A European Strategy for Sustainable, Competitive and Secure Energy – What is at stake? – Background Document, Commission Staff Working Document, Com(2006) 105 final, 8 March.

-------- (2006d), *Action Plan for Energy Efficiency: Realising the Potential.* Communication from the Commission, 19 October, COM2006(545) final.

European Council (2005), Presidency Conclusions, Brussels, 23 March, 7619/05 CONCL 1.

European Union (2003), Directive 2003/97/EC of the European Parliament and of the Council of 13 October 2003 establishing a scheme for greenhouse gas emissions allowance trading within the Community and amending Council Directive 96/61/EC, Official Journal of the European Union, L 275, 25 October, pp. 32-46.

-------- (2004), Directive 2004/101/EC of the European Parliament and of the Council of 27 October 2004 amending Directive 2003/87/EC establishing a scheme for greenhouse gas emission allowance trading within the Community, in respect of the Kyoto Protocol's project mechanisms. Official Journal of the European Union, L 338, 13 November, pp. 18-23.

ExxonMobil (2004), *Long-Range Economic and Energy Outlook* (see: http://www.exxonmobil.com/corporate/Citizenship/Corp_citizens hip_energy_outlook.asp).

Fujiwara, N., J. Núñez Ferrer and C. Egenhofer (2006), *The Political Economy of Environmental Taxation in the EU*, CEPS Working Document No. 245, Centre for European Policy Studies, Brussels.

Hoffert, M.I., K. Caldeira, G. Enford, D.R. Criswell, C. Green, H. Herzog, A.K. Jain, H.S. Kheshigi, K.S. Lackner, J.S. Lewis, H.D. Lightfoot, W. Manheimer, J.C. Mankins, M.E. Mauel, L.J. Perkins, M.E. Schlesinger, T. Volk and T.M.L. Wigley (2002), "Advanced Technology Paths to

Global Climate Stability: Energy for a Greenhouse Planet", *Science*, Vol. 298, 1 November, pp. 981-987.

IEA (2004), *World Energy Outlook 2004*, International Energy Agency, Paris.

IEA (2005), *Industrial competitiveness under the European Union emissions trading scheme*, International Energy Agency Information Paper, Paris.

IPPC (2001), *Third Assessment Report*, summary for policy-makers, Intergovernmental Panel on Climate Change, United Nations, New York.

OECD (1996), *Implementing strategies for environmental taxes*, OECD, Paris.

OECD (1997), *Environmental Taxes and Green Tax Reform*, OECD, Paris.

Pacala, S. and R. Socolow (2004), "Stablization Wedges: Solving the Climate Problem for the Next 50 Years with Current Technologies", *Science*, Vol. 305, 13 August, pp. 968-972.

Schellnhuber, H.J., W. Cramer, N. Kakicenovic, T. Wigley and G. Yohe (2006), *Avoiding Dangerous Climate Change*, Cambridge: Cambridge University Press.

United Nations (2004), *World Population Prospects: The 2004 Revision Population Database.* United Nations Population Division (see http://esa.un.org/unpp/).

WBCSD (2004), *Facts and Trends to 2050 – Energy and Climate Change*, report by the World Business Council for Sustainable Development, Geneva (http://www.wbcsd.ch).

PART II
TECHNICAL REPORT

J.C. JANSEN

AND

S.J.A. BAKKER

EXECUTIVE SUMMARY
OF TECHNICAL REPORT

In designing GHG reduction programmes, policy-makers to date appear to attach the most importance to cost-effectiveness, i.e. €/tCO₂ avoided. The application of this criterion for prioritising climate change mitigation options is highly problematic, however, due to:

1) widely divergent and partly mutually inconsistent practices in cost-benefit analysis (CBA), the paucity of data and large cost uncertainties; and

2) its disregard for many long-term social costs and benefits in which quantification problems constitute but one (important) underlying factor.

This report presents ancillary long-term social costs and benefits of CO₂ reduction and a proposed framework for their integration in cost-benefit analyses. On both the costs and benefits side, the estimates contained in the literature vary considerably. Key distinctive aspects include:

1) the perspective from which costs and benefits are (tacitly or explicitly) valuated,

2) the time horizon considered,

3) the rate at which costs and benefits are discounted,

4) the extent to which non-climate ancillary externalities are included in the analysis and

5) the uncertainties surrounding distinct costs and benefits.

Typically, most climate and ancillary benefits of GHG reduction activities can only be reaped after a relatively long 'gestation period', whereas the lion's share of the aggregate social costs typically needs to be absorbed soon after initiation of such activities. This inter-temporal asymmetry is closely linked with a key characteristic of the extent of sustainable development, i.e. the extent of inter-generational equity.

The focus of this report is on the design of a proper sustainability-oriented framework for the analysis of social costs and benefits of various climate change mitigation options. A second objective is to demonstrate such a framework by way of a numerical example to a selection of major climate change mitigation options in a European context. A key question of this study is: *How to bring externalities (ancillary costs and benefits) into mainstream practices of standard cost-benefit analysis?*

The essentials of the proposed standard framework for social cost-benefit analysis of various climate change mitigation options for public policy purposes are captured by the following broad guidelines:

1)   Check the interactions of the options reviewed and make sure that options retained for policy implementation purposes are not incompatible with each other.

2)   Use efficiency prices (i.e., by and large, market prices net of taxes and subsidies) as the point of departure for cost-benefit analysis from a societal point of view.

3)   Analyse explicitly the *context-specific* suitability of applied discount rates without 'automatically' applying discount rates used by authoritative economic development analysis and planning bodies.

4)   Show quantitatively the uncertainties surrounding the resulting key figures regarding mitigation costs per option.

5)   Make serious efforts to *quantitatively* include major external costs and benefits in the resulting key figures.

Starting out from a conventional framework, the proposed framework permits us to accurately and successively gauge the impact of alternative choices of the discount rate and distinct externalities on the resulting cost per CO_2-eq estimates. This is shown in the numerical example, which is based on the successive steps in the proposed framework diagrammed in Figure ES.1 below.

*Figure ES.1   Schematic overview of successive stages of the proposed framework
to arrive at net social cost projections of CO_2 abatement options*

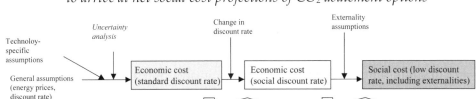

Two key externalities are explicitly addressed in the numerical example: i) impact on air pollution and ii) impact on long-term energy supply security risks. Regarding energy supply security, a novel approach is introduced to make due allowance for the impact of climate change mitigation measures on long-term supply security risk with regard to oil and natural gas. In principle, the approach can be readily extended to include coal and uranium as well.

Much attention is paid to uncertainty surrounding cost estimates of climate change mitigation efforts. Key factors impinging on final cost results turn out to be the choice of the discount rate and assumptions regarding future price evolution of fossil fuels oil and natural gas, and to a much lesser extent coal and uranium. In the numerical example, uncertainties amenable to the assignment of probabilities are brought out in the form of band widths (95% confidence intervals), based on Monte Carlo uncertainty analysis. It should be clearly stated that certain risks such as Damocles risks of major nuclear energy plant accidents and risks related to undesirable proliferation of nuclear technology and nuclear waste storage are not included in the results of the numerical example. In fact, it would appear that these ethical dilemmas are not amenable to quantification in any event but should be clearly included in any appraisal of the options concerned.

In social cost-benefit analysis, ideally a dynamic or rather an endogenous technology development approach should be used, where the cost of technology is not fixed and depends on other interacting technology developments as well as on policy dynamics. It is also important to acknowledge interaction between different options, not only on the physical impact of emissions reduction estimates, but on their mutual dependence as well. For example, development of CO_2 capture and storage (CCS) may depend to a certain extent on implementation of integrated gasification combined cycle (IGCC). Choices for certain technologies now may affect development of other options in the future. In our analysis, we have shown that implementing technologies 10 years later may change cost-effectiveness significantly, notably for wind and IGCC. However, this is only valid provided the assumed learning rates are achieved, which depend notably on the stimulation of the technology in its earlier years.

Based on the qualitative and quantitative assessment of the options studied, the following results:

- Insulation is very cost-effective (potential medium) from the end-user's point of view and has medium benefits for energy security and air quality.

- IGCC has medium costs but high air pollution avoidance benefits and contributes significantly to the longer-term goal of applying CCS, and to the development of cost-effective hydrogen production.

- Biofuels exhibit higher costs, but also offer high benefits for energy security.

- The costs of combined heat and power (CHP) are low to medium, depending strongly on future gas and electricity prices, but this option has medium ancillary benefits.

- Nuclear power appears to be cost-effective and has significant benefits regarding the avoidance of air pollution and energy supply security. Yet its suitability needs to be assessed in a much wider framework, including ethical issues not susceptible to quantification.

This study is of a relatively limited size. To carry out a genuinely comprehensive assessment would be a major project in its own right. In this context, the present report should be taken as a bold preliminary attempt to zoom in on the integration of ancillary long-term societal impacts in a socio-economic appraisal of climate change mitigation options. It is hoped, however, that this exploratory study may contribute to the future design of climate change policy in two ways, as outlined below.

1) The study provides a sustainability-oriented standard framework for social cost-benefit analysis by

    - making explicit allowance for long-term externalities;

    - showing that the inclusion of externalities in a quantitative fashion may offset the economic costs of climate change mitigation options outside the realm of climate change impact itself; and

2)   It represents a first effort to quantitatively account for long-term energy security of supply benefits.

Both contributions are meant to stimulate the debate in scientific and policy arenas. They will gain in significance if and when they are developed further. Using the proposed framework may also change CBA-based priority rankings evolving from the currently prevailing CBA approaches.

1. Introduction

Tackling climate change poses one of the world's greatest challenges. The evidence is getting stronger that most of the temperature rise that has occurred over the last 50 years is attributable to human activity. Authoritative international bodies, such as the IPCC (International Panel on Climate Change), suggest that gigantic global efforts are warranted to keep human-induced climate change within widely acceptable levels. The EU has identified climate change as one of the most important challenges it faces and has accordingly been engaged in a concerted effort to develop cost-effective policies for a coherent climate strategy. The process was kicked off by the European Commission's Communication of February 2005. This report articulates cost-effectiveness to be one of the leading criteria for the design of a European climate change programme (see European Commission, 2005a).

Mitigating human-induced climate change will, however, require a coherent and comprehensive long-term strategy, which places strong emphasis on cost-benefit analysis on a life-cycle basis. It is evident that priority should be given to those strategies that provide co-benefits in terms of economic efficiency, security of supply, containment of local pollution or innovation and job creation. Against this background, CEPS and ECN have undertaken a comparative cost-benefit analysis on different mitigation options. The objective of this project is to compare the different mitigation options on the basis of a social cost-benefit analysis with a view to informing the policy-making process.

This technical report proposes a sustainability-oriented framework for the analysis of social costs and benefits of different mitigation options on a life-cycle basis. It is based on a survey of the literature, complemented with spreadsheet model exercises that attempt to achieve a broad measure of comparability of the key cost-benefit results.

To date, cost-effectiveness of greenhouse gas (GHG) reduction options, i.e. €/tCO_2 avoided, appears to be the single-most important decision criterion for policy-makers in designing reduction programmes. It is important to note, however, that:

1)   an even-handed application of this criterion is difficult due to differences in both cost projection approaches and data availability; and

2)   in applying this criterion, many long-term social costs and benefits tend to be disregarded in which quantification problems constitute but one (important) underlying factor.

This report presents some details on ancillary long-term social costs and benefits of CO_2 reduction and a proposed framework for their integration in cost-benefit analyses. It reviews a selection of the recent literature on the cost and benefits of GHG reduction, with a focus on Europe. In consultation with CEPS, four main emitting sectors were selected for further consideration.

On both the costs and benefits side, estimates in the literature vary considerably. Key differences include:

1)   the perspective from which the costs and benefits are (tacitly or explicitly) valuated,

2)   the time horizon considered,

3)   the rate at which costs and benefits are discounted,

4)   the extent to which non-climate ancillary externalities are included in the analysis and

5)   the uncertainties surrounding distinct costs and benefits.

Most climate and ancillary benefits of GHG reduction activities can only be reaped after a relatively long 'gestation period', whereas the lion's share of the aggregate social costs typically needs to be absorbed soon after the initiation of such activities. This inter-temporal asymmetry is closely linked with a key characteristic of the extent of sustainable development, i.e. the extent of inter-generational equity.

It should be kept in mind that the scope of this study is relatively limited. A genuinely comprehensive assessment is a major project in its own right. In this context, given resource constraints, the present report should be taken as a bold preliminary attempt to zoom in on the integration of ancillary long-term societal impacts in the socio-economic appraisal of CO_2 mitigation policies and measures in a European context.

The remainder of this Technical Report is organised as follows. Chapter 2 reviews some general issues regarding the valuation of costs and benefits of CO_2 reduction activities from a societal point of view, leaving the externalities issue for more detailed treatment in chapter 4. A brief literature survey is made in chapter 3 of a selection of major reduction options in three sectors: i) energy and industry, ii) transport, and iii) residential and services. Chapter 4 zooms in on some of the main externalities that tend to be given little if any consideration of a quantitative nature in the standard literature on cost estimates of CO_2 reduction options. Results of the analysis of externalities are then used in chapter 5. This chapter presents the results of the application of the proposed analytical framework to a numerical example. It gauges the impact of some key externalities in (to the extent possible) a comprehensive social cost-effectiveness analysis compared to such an analysis from a more narrow economic efficiency perspective. The concluding chapter 6 presents findings and recommendations.

2. Valuation of social costs and benefits: Proposed methodology

2.1 Introduction

This chapter discusses some key methodological issues that have to be addressed in assessing the social costs and benefits of GHG reduction options. In doing so, costs and benefits will be considered from a broad, societal perspective as opposed to the narrow perspective of individual investors in greenhouse gas reduction activities. In principle, the societal perspective can be the perspective of a country, e.g. an EU member state, a region, e.g. the European Union, or the world. This report sets out to consider the perspective of the EU.

The chapter includes a brief review of recent literature on ancillary costs and benefits of the implementation of climate change policies and measures. Given the scope of the study, this review is by no means comprehensive. It rather aims at shedding light on the nature and the order of magnitude of major longer-term CO_2 emissions reduction impacts.

The following aspects will be reviewed:

- *Distinctive features of social cost-benefit analysis.* What makes it different from financial and economic cost-benefit analysis? How to compare the cost effectiveness between distinct GHG reduction options from a social perspective? (section 2.2)

- *Baseline setting.* Which reference situation will be used to assess the incremental costs and benefits of a GHG reduction activity? (section 2.3)

- *Discounting.* How to convert future costs and benefits to present values (section 2.4)?

- *Uncertainties.* Some major uncertainties for which the results of cost-benefit analyses tend to be rather sensitive are discussed. (section 2.5)

The main findings of this chapter, which result in five guidelines, are presented in the concluding section (section 2.6)

2.2   Social cost-benefit analysis: Key distinctive features

In considering the benefits and costs of specific GHG mitigation measures or projects, it is quite relevant from which point of view they are assessed and which system boundaries presumed. For example, are the cost and benefits resulting from proposed specific measures or investment projects considered from the perspective of their potential financiers or from the perspective of a national economy of a member state? What about impacts on other member states and the rest of the world? Without going into too much detail, we explain some key points that set social cost-benefit analysis apart from financial and economic cost benefit analysis.

Analysis of expected incremental[21] costs and benefits of a proposed project or measure will henceforth be referred to as *financial cost-benefit analysis* when only the perspective of financiers of a project is considered. This type of analysis takes prevailing and expected market prices (or, in other words, *financial prices*) as its point of departure, irrespective of whether these prices include indirect subsidies or taxes. The reason is that the overall (financial) return of a project to the financiers, taken together, depends on the evolution of market prices of resources (inputs) used. The investment-weighted average return to capital that potential financiers would expect as a minimum before agreeing to finance the project concerned, determines the *financial discount rate* by which expected incremental costs and benefits are converted to present values.[22]

Analysis of expected incremental costs and benefits of a proposed project or measure will henceforth be referred to as *economic cost-benefit analysis* when the impact is considered to the national (or regional) economy. To gauge the impact of a project on the national economy, market prices need to be adjusted for the impact of public sector intervention. Indi-

---

[21] Incremental cash flows are cash flows that can be attributed to the project, compared to a situation without implementation of the project concerned.

[22] We revert to the issue of discount rates in section 2.4.

rect subsidies received from the public sector in the member state concerned need to be added to market prices and indirect taxes removed. We then arrive at so-called *efficiency prices* or *economic prices*, which reflect the resource cost to the national economy of using the goods concerned.[23] In principle, economic analysis should correct for *external effects*, positive or negative effects upon the welfare of other individuals than the project financiers. In practice, to the extent that the positive or negative external effects are not readily quantifiable, these are often neglected in (narrowly-conceived) economic cost-benefit analysis. The *economic discount rate*, i.e. the discount rate applied in economic cost-benefit analysis of a project, is often taken to be the expected risk-free interest rate (the effective rate of interest on long-term government bonds) plus a premium reflecting the expected risk associated with the rate of future macroeconomic growth. Although project net benefit risk may co-vary less than perfectly with macroeconomic growth risk, in practice, a single 'official' discount rate is applied to the assessment of a broad variety of public investment proposals in a member state or region in line with recommendations of prominent national, regional or multilateral (economic) development agencies concerned. To avoid crowding out private capital by public capital, in setting the economic discount rate, due regard tends to be given to the average cost of capital in the private sector. In contrast, in conventional economic cost-benefit analyses, concerns for inter-generational equity tend to play a minor role in setting commonly applied discount rates.[3]

Analysis of incremental costs and benefits of a proposed project or measure will henceforth be referred to as *social cost-benefit analysis* when the impact of a project (or measure) to society at large is considered. In doing so, the system boundaries of the analysis, e.g. the member state concerned, the EU or the world at large, should be clearly defined. Social cost-benefit

---

[23] In principle, in deriving efficiency prices, allowance also needs to be made for non-tariff intervention, but this goes beyond the scope of this report. Where needed, further corrections need to be made through shadow pricing, when economic scarcities are not duly reflected (see e.g. the seminal handbooks for project appraisal: Squire & van der Tak, 1975 and Little & Mirrlees, 1975). In practice, such further adjustments are only made in the appraisal of projects in those developing countries that are characterised by very strong public interventions on the one hand as well as poorly functioning labour, capital, and foreign exchange markets on the other.

analysis of a project is based on the project's financial cost-benefit analysis. In financial analysis, however, market prices are used, while project impacts that do not have a financial bearing on the project financiers are neglected. Hence, for performing *social* cost-benefit analysis, the following adjustments are needed:

- Financial (market) prices need to be corrected for government interventions: indirect subsidies need to be added and indirect taxation subtracted.

- *External effects* need to be identified and their social value assessed to the maximum extent feasible (see chapter 4).

- A *social discount rate* – i.e. a rate of discount appropriate for social cost-benefit analysis of a project – needs to be applied instead of any financial discount rate that is only relevant to the project promoters (see section 2.5).

In principle, a project or measure can have a range of external effects. For instance, a GHG mitigation project setting out to achieve fuel switching from oil to biomass energy may result in significant benefits regarding mitigation of energy supply security risks. Substitution of natural gas-based electricity generation by wind power may even have significant local and regional air pollution reduction benefits. We will explain the issue of external effects in more detail in chapter 4.

To analyse public policy measures, a minimum, economic cost-benefit analysis needs to be applied but preferably social cost-benefit analysis would be conducted. Establishing the 'feasibility' of a measure from a (socio-) economic perspective, however, may require as a *condition sine qua non* that the measure should be feasible from a technical and financial point of view. For example, in assessing a measure targeting energy-efficiency improvement in industry by replacing a current manufacturing process by a more energy-efficient process, it should be established that:

i) the proposed production process is technologically mature, and

ii) including public interventions such as subsidies, it is financially attractive for the private sector to implement the proposed process.

Hence, taking another example, when assessing insulation measures targeted at households from a societal perspective, both the financial (end-user) perspective *and* the social perspective should be considered. When targeting home owners, in the case of rented houses so-called *split incentives* may complicate financial analysis. Take for example a home owner having to incur implementation expenses of, say, insulation while his tenants enjoy the financial return in terms of lower heating bills or, the other way around, tenants having to incur these costs while facing the prospect of moving within a short period of time. In such cases, the financial rate of return expected by the 'end-user', i.e. the targeted individual deciding on implementation of publicly incentivised energy-efficiency measures, might be very low indeed.

When a policy is targeted at achieving specific physical benefits of a kind that are hard to quantify in monetary terms, resort is often being taken to *cost-effectiveness analysis*. Then, the point of departure for prioritising policy measures from a societal perspective is either a given budget available for policy implementation or a given quantity of benefits to be achieved. Cost-effectiveness is then achieved by maximising the volume of targeted benefits for a given budget or minimising the required budget for achieving a given target for the policy objective concerned, i.e. CO_2 emissions reduction. As such, cost-effectiveness can be considered to be a special case of cost-benefit analysis, i.e. cost-benefit analysis per unit of targeted benefit. Hence, social cost-effectiveness analysis should also aim at allowing for external effects to the maximum extent possible.

In principle, most official (European) climate change policy documents adhere to the long-term policy objective to limit man-made warming of the earth to 2 °C by the year 2100 as a maximum. The problem is how to translate allowable global CO_2 equivalent emissions into a time trajectory and how much the EU or individual member states 'should' contribute. Given the complexity of this issue and the uncertainties regarding the post-Kyoto climate change regime, the best thing to do in supporting the design of climate change programmes appears to be to project CO_2 (or rather GHG) abatement curves[24] for the country or region concerned. Next policy-makers can prioritise CO_2 reduction options and determine the required budget allocation based on the CO_2 emissions (reduction) target or a certain maximum social CO_2 reduction cost per tCO_2 (tonne of carbon dioxide). The focus of this report is the methodology to be applied for assessing the

---

[24] Such curves depict the relationship between the estimated emissions reduction potential and marginal abatement costs in ascending cost order. In other words, the abatement (emissions reduction) potential of the cheapest option is shown first, then the next cheapest option, etc., in ascending order.

net social cost per tCO_2 abated when implementing a certain CO_2 reduction option.[25] This criterion provides quite useful information on the cost-effectiveness of different CO_2 reduction options, provided the options compared are not mutually inconsistent (see next section on interactions between options).[26]

When comparing the costs and benefits of distinct CO_2 reduction options, the way in which these values are summarised by means of a common yard stick is important. The total net cost can be either expressed as cost per tonne of CO_2 or cost per tonne of carbon. The difference between the two is a factor of 3.67, i.e. the difference between the molecular weights of CO_2 (44) and carbon (12). In terms of cost this would mean that a cost of €10/tCO_2 is the same as a cost of €36.7/tC. When assessing greenhouse gas abatement options, it is important to be aware of this distinction. An assessed value in terms of €/tCO_2 seems optically lower than a corresponding value expressed in €/tC. *In order to prevent confusion, we suggest the consistent use of cost of a CO_2 reduction option per tonne of CO_2 emissions avoided.*

A specific issue is how to discount the proposed *numéraire* for cost-effectiveness of a CO_2 reduction option: €/tCO_2. By applying the social discount rate (see section 2.5), the future net social costs of an option can be discounted. But how should we convert the value a tCO_2-eq of GHG emis-

---

[25] No attempt will be made to derive CO_2 abatement curves, as an estimation of CO_2 reduction potentials in e.g. the EU is far beyond the scope of this report.

[26] If a CO_2 emissions target could just be met by picking the low-hanging fruits of 'no regret' options that would be economically feasible even without considering their CO_2 reduction potential, another criterion should be used, i.e. maximisation of total net present value. However, if this situation was to exist in practise indeed, garnering political support for a global carbon emissions limitation agreement would not be so difficult. Stefan Thomas of the Wuppertal Institute suggests using two ranking criteria, i.e. the cost per MWh (including 'NegaMWh's of energy conserved) and emissions per MWh (Thomas, 2001). However, these criteria minimise the total cost of energy rather than necessarily leading to minimisation of net CO_2 reduction costs. We share the view that integrated resource planning (IRP) as such is a quite useful instrument in its own right. Notably·because of its integrated nature, IRP accounts well for interactions between distinct options. Yet this instrument cannot be applied *in isolation* to address more than one target (GHG abatement in addition to energy system cost minimisation).

sions one year from now to its corresponding value today?[27] If the global warming potential is similar and concentration levels of greenhouse gases tend to rise, why should the value of future GHG emissions be discounted anyway? Do future reductions matter less to mitigate human-induced climate change? There is no easy and fully satisfactory answer to these questions. Arguably, we suggest using a zero discount rate to convert future CO_2 reductions into present CO_2 reductions. In doing so, we presume that the actual impact on the climate change phenomenon per marginal unit of CO_2-eq emission is equal both over time and among individuals at any point in time.

2.3 Baseline setting for a social cost-benefit analysis

A crucial aspect in the assessment of incremental costs and benefits of a dedicated GHG reduction activity is the reference situation, i.e. the baseline against which the emission reductions and their associated costs and benefits are measured.[28] This is quite a tedious issue. A baseline is a counterfactual situation: when a proposed GHG reduction activity is implemented, the evolution of the state of the world in its absence is a matter of judgment that cannot be exactly verified in the real world. Hence, any determination of the baseline for whatever GHG reduction project is always susceptible to questions that cannot be resolved completely satisfactorily from a purely scientific perspective. Independent verification is crucial, however, as project investors and associated stakeholders may be tempted to inflate benefits in terms of GHG reductions and ancillary benefits. For example – even in the case of a country in which electricity generation is at present predominantly coal-based – a coal-based generation baseline should not be taken for granted in assessing the costs and benefits of a renewables-based

---

[27] This key question *should* be raised, even if a physical quantity cannot be readily equated to money or utility. For example, if it were to be established that an additional tCO_2 emissions next year would have less impact on climate change than a similar quantity emitted this year, or, alternatively assuming equal climate change impact that the value of consequential damages would be less for the tCO_2 emitted next year, a valid case could still be made in favour of using a positive discount rate. In practice, the issues of establishing the nature and size of climatic impacts and consequential damages are extremely complicated.

[28] See also Sijm et al. (2002).

generation project. Difficult questions need to be plausibly answered first. Will the proposed renewable power plant replace peak or off-peak power in the 'without project' situation? How will the future generation mix change to serve peak and off-peak demand? How will the heat rates (fuel efficiencies) of alternative power generation technologies evolve? A baseline scenario should provide plausible assumptions on, among others, such difficult questions.

The baseline should also properly account for recently implemented and officially announced or probable future policy changes. In the history of environmental regulation, tightening performance standards have kept on pushing the introduction of more environmentally-benign technological change. This phenomenon raises the question of whether or not a currently low emission technology can be considered additional (incremental). If this were not the case (i.e. equal emissions in the 'without project' situation because of a credible baseline that also shifts in an emissions-extensive direction), claiming incremental GHG reductions by way of projects patterned upon such a technology would not seem appropriate. For example, the European Commission (2003) points out that more efficient coal technologies (e.g. coal-based IGCC) is becoming cheaper due to technological learning. This may increase the GHG performance of coal-based generation compared to natural-gas-based generation. Hence, if – other factors remaining the same – IGCC will gain significance in the power generation sector, a shift from coal-based to natural-gas-based technology may result in less CO_2 reduction.

An important example of evolving policy frameworks is the recently designed European Directive on energy performance of buildings (European Commission, 2002). This directive promotes energy-efficiency measures and overall energy-efficiency standards for new and existing buildings. The key baseline question in this case is which measures will this directive mandate in any event and which will it others not mandate. Mandated measures are not additional and are therefore rejected for official qualification as GHG reduction measures.

An issue related to baseline-setting is *the possibility of interactions between GHG reduction options*. Reductions due to one specific measure may negatively impact the scope for reductions by another option. Examples include improving energy efficiency in power generation versus realising savings in final electricity consumption, building insulation versus more efficient heating or cooling systems, etc. These interactions should be mentioned and quantified where possible to give a realistic picture of costs and benefits. The occurrence of *'interdependent technology pathways'* may also be relevant, i.e. when the (scale of) implementation of one option may depend critically on another option. For example CCS will be much more attractive in IGCC plants compared to other coal-fired power plants. Investing in IGCC therefore partly determines the future of CCS. These dependencies are difficult to quantify, but need to be duly taken into account in the appraisal of individual GHG-reduction measures. This is certainly relevant for the design and implementation of coherent climate policy programmes as well.

2.4   Discount rate[29]

It is common practice to attach a higher positive (negative) value to a given unit of benefit (cost) realised today compared to one realised at some future date. Hence future benefits and costs attached to a GHG reduction option have to be discounted somehow to arrive at their present values. To do so, annual discount rates are used. If a cost-benefit analysis is carried out in real terms, future values are expressed in euros or dollars with the buying power of a certain point in time, e.g. euros of mid-year 2005. In that case, a real discount rate has to be applied, i.e. the discount rate after making allowance for projected general price inflation. Thus, if the *nominal discount rate* (including inflation) for a certain year would be 10% and price inflation 2%, the real discount rate would be approximately the difference, 8%. In the ensuing discussion, we will consistently assume the use of *real discount rates*.

The choice of discount rate used can impact highly on the relative attractiveness of a technology option. Compare for example a capital-extensive, expense-intensive generation option, e.g. natural-gas-based combined cycle gas turbine (CCGT) technology, with a capital-intensive, expense-extensive technology, e.g. onshore wind power. Asuming we use just one common discount rate, the choice of a 'high' rate, say in the 8-10% range, will favourably affect the expense-intensive technology, while the opposite holds for the choice of a 'low' rate, say in the 2-4% range. Choosing a 'high' rate will typically favour fossil-fuel-based, carbon-intensive

---

[29] See, among the many other publications on this subject, for instance Oxera (2002), Azar (2003), NEA/IEA (2005), Pearce & Turner (1990).

technology as future fossil-fuel expenses will be discounted rather strongly, whereas the opposite tends to hold for typically capital-intensive renewable options. This holds even when it is attempted to account for environmental externalities of fossil-fuel use, including notably GHG and polluting (SO_x, NO_x, particulate matter, etc.) emissions.

Yet, it is standard practice of international organisations such as the International Energy Agency (IEA) and the World Bank to use *one* discount rate to convert all future values into present values, irrespective of the nature of the technology, cost category or benefit category. Typically, a rather high rate on the order of 8-10% is opted for, compared to prevailing risk-free interest rates, which in OECD countries have been on the order of 1-3% (after inflation) in recent years. This practice is typically justified by making reference to the – more or less market-based – 'opportunity cost of capital' (OCC). In other words, a country should start out to invest its available capital resources in projects with the highest rate of return, then in the ones with the next highest return, etc. This process should be continued up to the point that all domestic capital resources are spent that are needed to bring about a fair distribution from a societal point of view between current consumption and savings for investment that enable future consumption. The return on the last unit of investment would coincide with the OCC. A soci(eta)al discount rate expresses among other things the societal premium attached to one marginal unit of current consumption over one marginal unit of future consumption against the backdrop of productive investment opportunities and the functioning of domestic financial markets.[30] Apart

---

[30] In theoretical treatises on social discount rates, reference is often made to the so-called 'Ramsey formula', dating as far back as 1928:

$$r = d + g\gamma$$

In this formula, **d** is the rate of pure time preference, **g** is the growth rate of the economy g and γ the elasticity of utility (indicating the extent to which social utility of income would be negatively affected by income growth). This formula indicates that many discount experts make a strong connection between growth and discounting. Typically investment opportunities in the developing world are riskier but – given good governance – entail higher profit prospects and hence opportunities for growth (technological leapfrogging, etc.) than in OECD countries. Moreover, capital markets in the developing world tend to function less well, making available capital resources scarcer. Therefore, in general it seems reasonable

from the consideration that, in practice, the values of the OCC and the social discount rate are hard to establish in a robust way, also from a theoretical perspective this practice is prone to contention.

Another issue related to the choice of discount rates is inter-generational equity in the face of the long-term trend towards deterioration of environmental qualities. Environmental conservationists tend to argue that sustainability and inter-generational equity would imply that "the" social discount rate should be revised downward to properly account for this negative trend, negatively affecting the well-being of future generations. This is a major concern leading quite a few (mostly environmental) economists to propose the application of discount rates that decline over time in a hyperbolic fashion (see e.g. OXERA (2002)). Discounting, say, large negative climate change impacts over very long timescales at high discount rates in fact boils down to virtually eliminating such impacts.

Indeed, sustainability considerations may lead policy-makers to somewhat reduce the general discount attached to future consumption in their revealed preferences. Yet, this is no remedy for failing to account for long-term externalities: when applying social cost-benefit appraisals to specific options, activities or projects, it is of utmost importance to make a proper effort in accounting *quantitatively* for environmental externalities *to the extent possible*.

Furthermore, the convention to simply use a one-size-fits-all discount rate for a wide range of different applications can be seriously challenged. Each GHG abatement option may have option-specific 'business risks'. For instance, financial cost-benefit analysis of electricity-generating investment projects stresses volumetric and short-term price risks on the output (commodity) market side. Yet from a societal point of view, electricity is a valuable commodity with relatively certain demand prospects compared to for example the social benefits of a large infrastructural project such as a new highway. Other factors being the same, it, therefore, seems reasonable to apply a comparatively lower social discount rate to the expected future benefit streams of power projects than infrastructure projects such as roads. The social discount rate of risky revenue streams should include an appropriate *societal* risk premium. Considering social cost streams from a societal

---

that in developing countries higher discount rates are used than in industrialised countries.

perspective, the natural gas expenses of a gas-based generating technology, e.g. CCGT, are quite risky and typically far less than perfectly (positively) correlated to economic business cycles. Conversely, if the CCGT plant will be used in base/intermediate load mode, the future investment costs of such technology are much less uncertain. This would prompt the use of a relatively low social discount rate for the fossil-fuel cost streams compared to the capital expenditure. Hence, *theoretically* a 'one size fits all' discount rate over time for each and every technology and cost category is less appropriate.

On the other hand, any differentiation in the discount rate applied between distinct future time periods, technologies and/or cost categories would be rather controversial. Moreover, it would need very detailed location-specific information. *Weighing these and earlier mentioned considerations, we suggest the consistent use of a single social discount factor with a fairly moderate value as compared to standard practices of leading international agencies, say in the 3-5% range with 4% as central value.*

2.5   Uncertainties

Cost-benefit analysis of GHG emissions reduction options is surrounded by huge uncertainties. Consider for example the uncertainties regarding technological developments, especially unexpected fossil-fuel-saving innovations. In the following section, we discuss uncertainties regarding one other key uncertainty: the future price trajectories of fossil fuels.

Recent price projections for year 2030, which are often used as a reference regarding 'the' world oil price, are as follows: the most authoritative projections on the part of the European Commission are those commissioned by DG TREN and prepared by NTUA (e.g., European Commission, 2003). In the baseline scenario of the aforementioned publication it is stated:

> ...no supply constraints are likely to be experienced over the next 30 years ...Oil prices decline from their high 2000 levels over the next few years, but they then gradually increase to reach a level in 2030 no higher than that in 2000 (and 1990)...

In this baseline scenario the crude oil price is expected to move from USD_{2000} 28.0/boe in year 2000 to USD_{2000} 27.9/boe in year 2030. In 2004 this projection was revised upward by a considerable margin, i.e. USD_{2000}

28/boe (low price scenario) to USD$_{2000}$ 50/boe (high price scenario).[31] Given intervening real market developments a substantial further upward revision in the next update would not come as a surprise. Also the IEA has appreciably revised upward its world oil price projections for 2030 in the latest (2005) World Energy Outlook with a price of $$_{2004}$ 39 in its reference scenario and $$_{2004}$ 52 in its alternative scenario, against a price of $29 in the Reference Scenario of the 2004 World Energy Outlook issue. The Overview of the 2006 Annual Energy Outlook (AEO) of EIA (DoE, 2006: 1) states:

> ...In the *AEO2006* reference case, world oil prices, which are now expressed in terms of the average price of imported low-sulphur crude oil to U.S. refiners, are projected to increase from 40.49 per barrel (2004 dollars) in 2004 to $54.08 per barrel in 2025 (about $21 per barrel higher than the projected 2025 price in *AEO2005*) and to $56.97 per barrel in 2030...

In addition to the issues in the previous section, uncertainties in parameters play an important role in cost (and benefits) assessments. Key variables that determine cost of GHG reduction are energy prices. The years 2004 and 2005 have displayed a strong increase in oil, gas and electricity prices, both globally and in Europe. Whether these prices are only a short-term price hike or whether in 2020 the cost of energy will be equal to or higher than current levels is a major question that impacts on the cost of energy-related GHG mitigation options.

Table 2.1   Energy price projections for year 2030

energy carrier	unit	WEO 2005 (IEA, 2005)		Primes 2004[a] (EC, 2004)		AEO 2006 (DOE, 2006)	
		Reference	High	Baseline	High	Reference	High
oil (average crude IEA import)	$/2004/bl	39	52	28	50	57	95
	$2004/GJ	6.8	9.1	4.9	8.8	9.9	17
gas European imports	$2004/Mbtu	5.6	7.1	4.3	7.7	6.9	9.0
	$2004/GJ	5.3	6.8	4.1	7.3	6.6	8.6
coal (OECD steam coal imports)	$2004/tonne	51	57			31	46
	$2004/GJ	1.8	2.0			1.1	1.6

a) $2000/GJ.

b) AEO gas and coal prices are US domestic.

So far, in many cost-benefit analyses of CO_2 reduction technology options, a stable level of fossil fuel prices is assumed. In doing so, cost-benefit

---

[31] European Commission, 2004.

analysts usually refer to price projections published by reputed bodies, such as the IEA, EIA (European Information Administration) or the World Bank. In this way, individual analysts may avoid potential blame of assumed *ex ante* price trajectories that turn out to be widely out of line with *ex post* price development.

However, international bodies have a notoriously poor track record in predicting the future price of oil and natural gas. Typically, these organisations predict stable or gently increasing price evolution trajectories starting out from the most recent available price data. The high year-to-year volatility in real world oil price developments is mirrored to a large extent by year-to-year shifts in projected price trajectories by the IEA and other often-quoted publications. A more recent trend for the IEA and other providers of official energy price projections is to use several scenarios instead of just a reference scenario accompanied by sensitivity analysis with regard to fossil fuel prices. Nonetheless, with regard to reference scenarios, there seems to be a strong tendency to underestimate future price increases of fossil fuels (see e.g. Bolinger et al., 2004).

In general, uncertainties surrounding the factors that determine the cost of GHG mitigation options should be clearly presented. In the numerical example of the social cost-benefit analysis methodology proposed in this report, we will reflect uncertainty regarding this value of underlying factors in bandwidth estimates of the reduction cost per tonne CO_2-eq for selected options through the application of Monte Carlo uncertainty propagation analysis using special @RISK software. This method takes as its point of departure the assumed expected bandwidths (95% confidence intervals) and central values of underlying cost factors. Through calculation simulations, it arrives at central values and 95% confidence intervals for the abatement cost of the options considered (see also Annex A.1).

2.6   Summary of findings

Credible baseline determination for comparative social cost-benefit analysis of distinct emissions reduction options is far from simple and implies a fair amount of subjective 'expert judgment'. This relates to the fact that the 'without project (programme)' situation is counterfactual.

Difficulties include the dynamic nature of a credible baseline and interdependencies between different technological options.

Hence, conducting a proper procedure for expressing future costs and benefits in present-day values is far from easy. Arguments have been presented that prominent international bodies such as the IEA tend to use a 'one-size-fits-all' social discount rate that would seem on the high side and thus discriminating in favour of capital-extensive, expense-intensive projects. Notably fossil-fuel-based options tend to be expense-intensive. In the context of this report, we propose to use a real (i.e. excluding general price inflation) discount rate with a more moderate risk premium over the risk-free rate. In EU capital markets, a risk-free real interest rate on the order of 1-3% has been realised over the last decade. Considering the above, using a social discount rate on the order of 3-5% is considered reasonable.

As the measure for cost-effectiveness of an option, we propose to use the quotient of discounted total net cost and undiscounted total CO_2 emissions reduced.

As evidenced by the previous findings, cost-benefit analyses of GHG reduction options are surrounded by huge uncertainties. So are the baseline assumptions on future price trajectories of oil and natural gas. There seems to be a tendency by providers of 'official' price projections (such as the IEA, World Bank, EIA) to seriously underestimate future price increases and volatility. Therefore, prudence in using these projections seems in order.

For applying social cost-benefit analysis to distinct climate change mitigation options for public policy purposes, we suggest heeding following broad guidelines:

1. Check the interactions of the options reviewed and make sure that options retained for policy implementation purposes are not incompatible with each other.

2. Use efficiency prices (i.e., by and large, market prices net of taxes and subsidies) as the point of departure for cost-benefit analysis from a societal point of view.

3. Analyse explicitly the *context-specific* suitability of applied discount rates without 'automatically' applying discount rates used by authoritative economic development analysis and planning bodies.

4. Show quantitatively uncertainties surrounding resulting key figures regarding mitigation cost per option.

5. Make serious efforts to *quantitatively* include major external costs and benefits in resulting key figures.

3. GHG reduction cost for selected options: A brief survey

This chapter presents a brief survey of the literature on the cost of greenhouse gas emissions avoided per tonne of CO_2 equivalent in European countries and regions. Given the limited size of this study, the survey is far from comprehensive. We will focus the analysis on a selection of major technology options for GHG emissions reduction. Points of interest are the methodologies applied and the resulting cost estimates. An explanation of the technology selection is given in section 3.1. In the ensuing sections, options are discussed in greater detail in the sectors energy and industry (section 3.2), transport (3.3) and residential and services (3.4).

3.1 Selecting the GHG emissions reduction options

A broad overview of sectoral contributions to CO_2 emissions in the EU is given in Table 3.1 and Figure 3.1. Total annual CO_2 emissions in the EU add up to 3.7 $GtCO_2$ (year 2000; excluding sinks and agriculture). Note that only direct emissions are given for each sector, so that emissions from e.g. electricity consumption are not included.

Table 3.1   EU-25 CO_2 emissions: level (year 2000) and average annual growth (1995- 2000)

Sector	CO_2 emissions year 2000 (Mt)	Average annual growth, 1995-2000 (%/yr)
Electricity and steam production	1228	-0.2%
Energy production/conversion, n.e.s.	164	0.0%
Industry	606	-1.2%
Residential	463	-1.1%
Services	237	-1.2%
Transport	968	2.4%

n.e.s = Not elsewhere specified.

Source: European Commission (2004).

Figure 3.1 CO₂ emissions in the EU-25 in 2000

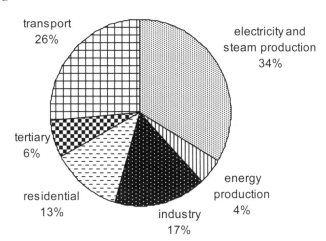

Source: European Commission (2004).

Important considerations for policy-makers to select GHG reduction options for inclusion in their climate change programmes (if applicable) include:

- cost-effectiveness (expressed in €/tCO₂-eq)
- co-benefits for other policy areas such as energy security
- certainty about cost and benefits
- GHG abatement potential
- public acceptability
- ease of implementation
- no major negative and preferably positive interactions with related options.

Respecting these criteria, we make a broad classification of major reduction options in each main sector. The first classification level refers to the unit cost level (cost per tCO₂ reduction). Two major classes are distinguished: 'no regrets/low' and 'medium/high'. These classes can be roughly equated to 'unit cost ≤ €20/tCO₂' and 'unit cost > €20/tCO₂'. The second classification level refers to the nature of the net ancillary benefits.

Two sub-classes are distinguished: one containing options with unambiguous and notable net ancillary benefits, labelled *'clear benefits'* and another one containing options with more ambiguous ancillary benefits and/or major implementation problems, labelled *'implementation problems'*. Our resulting classification of options is shown in Table 3.2. This table serves as a first classification of major options.

Table 3.2   Preliminary classification of GHG mitigation options

Cost	Ancillary benefits/problems[a]	Sector[b]	Options
No-regret low	Clear benefits	EI	Energy-efficiency (DSM); low-clinker cement
		EI	CHP; Biomass co-firing
		RS	Insulation; Efficient heating; lighting
		T	Fuel switch
No-regret low	Implementation problems	EI	Recycling; alternative fuel in cement industry
		EI	Nuclear
		RS	More efficient heat generation
		T	Modal shift; traffic management; Fuel economy
Medium high	Clear benefits	EI	Waste heat utilisation
		EI	RES-E (wind, biomass)
		RS	High-efficiency heat generation; Solar power/thermal
		T	Biofuels, Hybrid vehicles
Medium high	Implementation problems	EI	CCS[c]
		RS	Micro-CHP
		T	Hydrogen fuel cells

[a] Including 'certainty of benefits and problems'.

[b] EI: energy and industry, RS: residential and services, T: transport.

[c] Synergy with development of H_2 generation is a possibility.

Source: Authors.

Given the resource constraints of this explorative study, we have retained a 'shortlist' of selected options for further analysis. This shortlist is based on 1) the pragmatic consideration of data availability, 2) coverage of the most important sectors and 3) both short-term and long-term options with a significant potential. The analysis is also focused on energy-related CO_2 emissions rather than non-CO_2 GHGs. The shortlist encompasses both the mitigation options selected and selected reference options. The latter serve as baseline.[32] The shortlist encompasses:

1) Energy and industry
- Wind on-shore
- IGCC (instead of PCC)[33]
- Biomass (co-firing in coal plant)
- Nuclear
- Combined heat and power (natural-gas-based CCGT)
- CO_2 capture and storage PCC plant

With reference options:
- PCC (coal-based)
- Natural-gas-based CCGT

2) Transport
- Future biofuel (cellulose-based)
- Hydrogen (CO_2 low or neutral)

With reference option:
- gasoline

3) Residential and service sectors
- Insulation residential/commercial
- Efficient heating installations

---

[32] In the power sector, the quantity of CO_2 emissions reductions per unit, say per MWh, of the different mitigation options depends on the carbon intensity of the marginal generating technology. For example, CO_2 reduction from wind power will be higher if it replaces power from coal compared to the case that it would replace electricity from an IGCC (integrated gasification combined cycle) power plant. Hence, the choice of reference option is quite relevant. Section 2.3 discusses baseline setting in greater detail.

[33] For the meaning of abbreviations: see the glossary of abbreviations.

With reference option:

* Pre-insulated residential/commercial

3.2   Indicated cost of selected options in energy and industry

Most of the selected options in the energy and industry sectors refer to power generation rather than to other energy-related emissions. Current installed capacity in the EU-25 power sector amounts to 650 GW$_e$. With demand projected to increase to approximately 730 GW$_e$ in 2020 (assuming no change in the overall capacity factor) and the need to replace fossil and nuclear capacity older than 40 years, more than 300 GW$_e$ needs to be built by 2020 (IEA, OECD, 2005b). This figure and long capital turnover periods render choices made now regarding electricity and climate policy to have a large and long-term impact.

Let us summarise some general results from two recent studies. NEA/IEA recently published projected costs of electricity from various sources: coal, gas, nuclear and wind (NEA/IEA, 2005). The objective of this study was to inform policy-makers about the economic cost of generating electricity, and the fact that externalities are not taken into account. We used some data from this source to assess electricity generation costs and GHG abatement costs, using varying baseline assumptions and energy prices (see chapter 5). Table 3.3 shows results of the study for the EU-15 countries. These are figures calculated using 5% and 10% discount rate. Comparing figures with other calculations should be done with care, as outcomes are very sensitive to assumptions.

Table 3.3   Electricity generation cost according to NEA/IEA

Fuel	Cost (US$/MWh)		Major assumptions / remarks
	5%	10%	Discount rate
Nat. gas	38-56	41-59	Gas price $ 3.72-6.65 /GJ 2010
Coal	22-48	28-59	Investment cost $400-1000/kW$_e$
Nuclear	23-36	32-53	Investment cost $1100-2100/kW$_e$
Wind	35-95	45-120	Inv. cost $1000-1900/kW$_e$ (mostly on-shore)

Source: NEA/IEA (2005).

The figures for gas and coal-based electricity can be compared to cost estimates by IIASA (2005),[34] which amount to €44 and €38/MWh respectively $53 and $46/MWh). On the other hand, IIASA's estimate for nuclear is higher than NEA/IEA's – €44/MWh ($53/MWh) – and for on-shore wind lower: €42/MWh ($50/MWh).

The World Alliance for Decentralised Energy has published a critical review of the IEA/NEA study. Its main points are: 1) the difference between generation and delivered cost is of utmost importance to the consumer, 2) assumptions on the economic lifetime and capital cost in the study unreasonably favour nuclear and coal plants, 3) high capital cost plants are very sensitive to the discount rate applied, and 4) the methodology does not take into account business risk associated with competitive energy markets (WADE, 2005).

Survey highlights for each of the selected power generating options are presented in the ensuing sub-sections.

3.2.1 Wind power

The potential for wind-based power generation is mainly limited by wind availability as well as by geographical and spatial planning constraints. As shown by the case of Denmark, a high penetration of wind power can be reached when attractive incentives are in place. The initial investment is the main part of the generation cost, while the variable and fixed operating and maintenance cost are relatively low. Therefore, the choice of discount rate has a high impact on wind power cost calculations.

According to Hoogwijk (2003), the global potential for wind on-shore is approximately 20 PWh/yr with a generation cost of between €50 and €70/MWh. This study assesses generating costs using a 10% interest rate and 20 years lifetime. For the European Union, IIASA (2005) states an economic potential of about 300 TWh/yr in 2020 (on- and off-shore), which corresponds to approximately 110 or 250 $MtCO_2$/yr reduction, if gas-based or coal-based electricity is replaced. In the Danish 4th National Communication on Climate Change, €35/tCO_2 socio-economic abatement cost is re-

---

[34] These cost estimates represent the values to society of allocating resources to reduce air emissions. In the calculations, a 4% discount rate is used.

ported for on-shore wind energy in 1990-2001 (Danish Ministry of Environment, 2005).

3.2.2 IGCC, coal-based

Integrated gasification combined-cycle power plants achieve a higher efficiency and thus lower CO_2 emissions per unit power production (approximately 20% lower) as well as a substantial reduction in NO_x, SO_2 and particulate matter, as compared to pulverised coal combustion (Lako, 2004). Only a few IGCC plants are currently operating, and it is still seen as an immature technology. As it is generally anticipated that coal-based power production will be important in the coming decades, IGCC is a crucial technology for more sustainable power production (Van der Zwaan, 2005).

The current investment cost of an IGCC plant is about €1700/kW$_e$, but is expected to decrease significantly in the coming years. It is anticipated that conversion efficiency will improve from 45% currently to 54-56% in 2020, possibly even higher. IGCC fuels can be coal, residuals and biomass (Lako, 2004). Applying CO_2 capture and storage in such a plant is also much more cost-effective than implementing this in a PCC plant. Therefore, IGCC can play an important role in stimulating CCS in the medium term. Also IGCC can play an important role in the development of hydrogen energy systems, providing a possible low-cost source of hydrogen.

For Germany, IEA/OECD (2005b) estimates CO_2 abatement cost for new IGCC plants to be €20-29/tCO_2, assuming a new plant of 1.05 GW is built. The report also acknowledges the possible important role in promoting cleaner and CO_2-free coal-based power generation.

3.2.3 Biomass co-firing

Biomass can be used as a fuel for co-firing in coal-fired power plants. There is little loss in combustion efficiency when burning 10% biomass (Smith, 2001). The greatest potential for biomass is in plants operating with circulating fluidised bed coal combustion, where 50% biomass can be co-fired.

This requires an investment in the hardware of the power plant, such a separate grinding and firing equipment, which CPB/ECN (2005) estimates to cost €600/kW$_e$ for the Netherlands. A 7% discount rate is used in evaluating alternative policy options for CO_2 reduction, where an attempt is made to include external effects on air pollution and some aspects of energy supply security (i.e. intermittency). Average biomass fuel cost is estimated at €48/MWh. Several types of biomass can be co-fired in this fashion, including agricultural waste, wood chips and bio-oil.

Hoogwijk (2003) estimates the global technical biomass potential in 2050 to be 60 PWh/yr, at generation cost of €45/MWh, in which case this option would be competitive with fossil options at low CO_2 prices. This is based on availability of biomass $2/GJ_p$, especially from the former USSR, Oceania, East/West Africa and East Asia. According to IIASA (2005), generation cost are higher, more than €70/MWh, at 2020 prices of €3-5/GJ. The economic potential would then be approximately 300 TWh/yr in the EU-25, or 200-250 $MtCO_2$.

ECN/MNP (2005a) estimates co-firing to cost approximately €52/tCO_2 in 2020 in Dutch coal-fired power stations. This document aims to evaluate the policy effectiveness for GHG emissions reduction options in the Netherlands and uses an economic approach with a 4% discount rate. The analysis does not include external effects.

3.2.4 Nuclear power

Nuclear power is characterised by high capital costs (approximately €1,900/kW_e, although estimates vary considerably) and lower fixed and variable operation and maintenance costs, compared to fossil-fuels-based generation. Calculations of electricity commodity prices depend mainly on the investment cost per unit of capacity and the discount rate. According to the NEA/IEA, nuclear power is in many cases more cost-effective than gas or coal-based power generation, making this a potentially important GHG mitigation option. This is confirmed by ECN/MNP (2005a) for CO_2 options for the Netherlands, in which the mitigation cost of nuclear power is estimated to be €8/tCO_2, where the storage cost for nuclear waste for 100 years, and the insurance cost for nuclear accidents are taken into account.

An important issue with nuclear power cost estimates is the extent to which end-of-life capital costs for decommissioning are taken into account. The World Alliance for Decentralised Energy (WADE, 2005) assumes in its electricity cost calculations a set-aside cost for decommissioning of $2,5000/kWe, spread over 40 years. This would amount to approximately $7/MWh if spread undiscounted over the 40-year power production.

In the ExternE project, it is calculated that the (external) cost associated with nuclear reactor accidents is relatively small. Using a certain monetary evaluation of human life, this cost is estimated to be less than 1 euro cent 1 per MWh. However, it is also acknowledged that this valuation methodology is not suitable for so-called 'Damocles risks' – very low probability risks of occurrence of events with a very high damage (ExternE, 2005). Another important question is how to account for nuclear waste disposal costs, including the time horizon. A recent survey in 18 countries around the globe, including UK, France, Germany and Hungary, showed that most citizens still oppose construction of new nuclear power stations. Even when the climate-change benefits were highlighted, only a small number of respondents became more supportive (IAEA, 2005).

The availability of fuel appears not to be limited in the immediate decades to come, but in the longer term (beyond 50 years) uranium availability may become a problem (NERAC, 2002). Therefore the technical potential of nuclear power is not an important limiting issue. On the economics side, the availability of sufficient financing and the rather long lead-time (5-10 years) are more important. Ultimately, using more nuclear power is more of a social and political choice, where the different types of risks, inter-generational aspects and the short- and long-term environmental aspects have to be weighed.

3.2.5   CO₂ capture and storage

CO_2 capture is the most costly step of CCS (CO_2.capture and storage). It can be applied in power production in three different systems:

- Post-combustion, in which CO_2 is separated from the exhaust stream after fuel combustion using e.g. membrane separation;

- Oxy-fuel combustion, using high-concentration oxygen in order to produce a more pure CO_2 stream, reducing the separation cost (but energy to produce oxygen is required); and

- Pre-combustion, where the fuel is gasified into H_2 and CO_2 (in two steps) and the hydrogen is used for power production.

The 'energy penalty' to capture the CO_2 is generally highest in post-combustion. However, as this technology is commercially available, in contrast to the other two, and therefore the most likely to be applied in the mid-term, we present figures only for this technology.

After capture, the CO_2 needs to be compressed and transported by ship or pipeline to the storage site (e.g. empty gas field or saline aquifer) and injected into the geological reservoir. (The International Institute for Applied Systems Analysis IIASA, 2005) estimates these costs to be €8-24/tCO_2 stored.

The IPCC (2005) *Special report on CO_2 capture and storage* reviewed cost studies on CCS technology and the reference technology with the aim to provide policy-makers with state-of-the-art knowledge. It is an economic analysis, but the assumptions and approaches used by the reviewed studies were not mentioned explicitly. The report estimates \$30-70/tCO_2 avoided for total system cost when a pulverised coal plant is compared to one with CCS (IPCC, 2005). When a natural gas combined cycle plant or an IGCC (integrated gasification combined cycle) is equipped with CCS, costs are comparable, using a pulverised coal plant as a reference. System costs may be lower if applied in Enhanced Oil Recovery, in which the CO_2 is used to extract oil from oil fields, which would have been uneconomical otherwise. For CCS implemented in new combined heat and power plants in the Netherlands, ECN/MNP (2005a) estimate the CO_2 reduction cost to be €56/tCO_2 avoided.

3.2.6 Combined heat and power in industry

The significant role CHP can play in climate change mitigation and energy security has been recognised in the EU with the Directive on promotion of cogeneration based on a useful heat demand in the internal energy market (2004/8/C). It sets a common framework for calculating energy efficiency gains and requires member states to report on installed capacity, but does not impose mandatory targets.

Integrated production of heat and power achieves a considerably higher primary energy conversion compared to separate generation. Many industry sectors require a large input of heat (often in the form of steam) for their production processes, e.g. pulp and paper, petrochemical and food and drug industries. In CHP, part of the generated steam is used for power production, while the rest is transported to the industrial production site to be used as process steam. Different technologies exist to meet the energy needs of the affected industry. The most important variable is the ratio of heat and power (as expressed in MW_{th}/MW_e), and different technologies exist that have different characteristics. A barrier for implementation of CHP is the difference in (diurnal) demand patterns for heat and power and therefore it should be possible to operate such capacity in a flexible manner. It can be implemented in two ways: an industry covering its heat demand and using or selling the generated electricity, or a joint venture between a power company (using most of the electricity) and an industry (using the heat).

Smith (2001) states that there is potential to use CHP on coal-fired generation capacity in most countries, and that it is one of the most cost-effective CO_2 reduction measures, both for coal and gas-fired units. It may save up to one-third of the fuel used when compared to separate power and heat generation.

The cost-effectiveness of CHP depends mainly on the difference in prices of primary fuel (gas, coal) and electricity. ECN/MNP (2005) deem process-integrated CHP in petrochemical production sites an option with a low cost (€0 to 10/tCO_2). Similar figures apply to large-scale CHP potential in other industry sectors.

3.2.7 CCS in refineries, fertiliser, natural gas production

Cost-effectiveness of CO_2 capture and storage depends to a large extent on the scale of CO_2 emissions and therefore any large point source can be eligible for application of the technology. In addition to power production, energy-intensive industries should be considered. Due to process characteristics, fertiliser, refineries and hydrogen production are likely to be the cheapest options.

IPCC (2005) recently published a comprehensive report on different aspects of CO_2 capture and storage, including cost assessments. These are based on current costs, while it also estimates that a cost-reduction potential of 20-30% for capture is possible. Assumed prices are $15-20/bll of oil, $2.8-4.4/GJ for gas, and $1-1.5/GJ for coal. At higher fuel prices, abatement costs will increase due to the inherent 'energy penalty' involved in CO_2 capture. This penalty ranges currently between 14-20% with existing technologies and is forecasted to be around 7-17% by 2012 (Ha-Duong & Keith, 2003).

Table 3.4   CO$_2$ capture and storage cost in industry ($/tCO$_2$ avoided)

	$/tCO$_2$	
CO_2 capture from hydrogen and ammonia production	5 - 55	(-5) – 30 with EOR
CO_2 capture from other industrial sources	25 - 115	
Transportation	1 - 8	
Geological storage + monitoring	0.5 - 8	

Note: Figures cannot be simply summed to calculated system costs.
Source: IPCC (2005).

According to ECN/MNP (2005a), the abatement cost for CO_2 capture and storage in Dutch refineries, ethylene production and ammonia production is approximately €8-10/tCO_2.

An early opportunity for the application of CCS is in the process of natural gas production, also called 'gas sweetening'. CO_2 separation is necessary for natural gas transport specifications and therefore only transportation and storage costs have to be taken into account when calculating the CO_2 abatement cost. This is already implemented in Norway and Algeria. It is estimated that CCS in gas sweetening can be carried out at prices lower than €30/tCO_2 avoided, as shown by the Sleipner project in Norway, where 1 MtCO_2 annually has been injected since 1996 at cost of €18/tCO_2 (Torp et al., 2004).

3.3   Indicated cost of selected options in transport

The transport sector appears to be a difficult sector to address with climate policy. CO_2 emissions from this sector are rising rapidly, both from land and air transport. Policies to slow the increase in emissions are relatively ineffective, in the short term, but while it is sometimes suggested that such policies have been effective in the longer term. Here we discuss two of the main options regarding alternative fuels: biofuel and hydrogen.

The choice for these two options does not mean that other measures, such as fuel economy improvement, are less important or not cost effective. IIASA (2005) assumes in its GAINS model a 25% reduction in specific CO_2 emissions for the improved gasoline car, and a further reduction for hybrid models.

3.3.1   Future biofuels

Fuel from energy crops can be used to partially substitute gasoline or diesel, requiring no additional investment in engine technology. For high blending percentages investments into alternative engine materials may be necessary; nevertheless the additional cost per vehicle is likely to be modest (IEA, 2004). An important distinction can be made between conventional biofuels (e.g. pure vegetable oil, biodiesel and ethanol) and so-called future biofuels: ligno-cellulose based bioethanol, Fischer-Tropsch diesel and bio-dimethylether. Currently, conventional biofuels are cheaper than future biofuels, but Wakker et al. (2005) expect that from 2010 the market share of the latter will increase. CO_2 reduction compared to fossil fuel, per unit of energy, is much higher for the future biofuels – 90% versus 45% for conventional biofuels. CO_2 reduction of biofuel depends on production source, transportation distance and fuel processing. It can be estimated that €8/GJ for future biofuels is possible, provided that advanced biofuel processing technologies develop according to expectation, and that the cheap woody biomass potential in Central and Eastern European countries becomes and remains available at a price of €1.5-3/GJ. In this case future biofuel will be able to compete with oil at prices of $60-100 per barrel in 2010. The total biomass potential in 2030 is very large: over 10 EJ/yr, and 20% biofuel in the transportation appears realistic under favourable scenarios (Wakker et al., 2005). One constraint is availability of land. If 10% of the land currently used for agricultural purposes in the EU is dedicated to biofuel production, 8% of current gasoline and diesel consumption can be replaced (IEA, 2003).

The feedstock for biofuels competes with other biomass options, such as biomass in power production, and land for agricultural purposes. There-fore, the feedstock price is surrounded by high uncertainty. The biofuels option also interacts with efficiency gains in vehicles and alternative fuel options such as natural gas and hydrogen fuel cells.

Few abatement cost estimates have been found for future biofuels. Es-timates will also be highly uncertain, due to uncertain feedstock costs in the future, N_2O emissions from agriculture and future oil prices. It will also depend on the biomass yield per hectare, which varies across geographic regions. IEA (2004b) projects €60-140/tCO_2 for cellulosic ethanol after 2010 (as compared to €150-210/tCO_2 currently, in the pre-commercial stage). For comparison, we note that the same report calculates €200-500/tCO_2 for cur-rent grain-based ethanol and €10-60/tCO_2 for cane-based ethanol in Brazil. The study makes an economic analysis based on efficiency prices and re-views state-of-the-art knowledge in different world regions. It addresses external costs and benefits separate from the cost calculations, noting they can be substantial even if it is not possible to quantify them.

3.3.2 Hydrogen fuel cells

Hydrogen is often touted as *the* fuel of the future, for reasons of reducing dependency on oil and improving air quality. According to IEA/OECD (2005a), a transportation system based on hydrogen fuel cell vehicles may – depending on the well-to-wheel energy chain-result in a substantial reduc-tion in oil demand and primary energy use. Currently large research budg-ets are allocated to promote demonstration projects and larger-scale com-mercialisation of both fuel cell (or hybrid) vehicles and hydrogen production. Hydrogen production from coal with CO_2 capture and storage would also help in combating climate change.

Nevertheless, large utilisation is a major challenge and appears to be nothing less than a gigantic transition in the transport energy system. This transition requires large investments in new types of vehicles, fuel production and, most significantly, infrastructure – and appropriate policies. It is therefore very difficult to give an estimate with reasonable certainty on the GHG reduction cost of the entire system.

Cost for both the fuel and the fuel cells are substantially higher compared to the alternatives. According to a recent report by the IEA/OECD (2005b):

> Depending on the production technology, hydrogen production cost should be reduced by a factor of 3 to 10, while the cost of fuel cells needs to be reduced by at least a factor of 10 to 50, in comparison with current cost estimates. Technology learning is the key to achieving these targets.

IIASA (2005) considers 2% as a maximum market penetration of hydrogen fuel in the passenger car fleet in 2020 and €17/GJ fuel price, assuming it is produced from fossil fuels with CCS. This production method appears to be the most likely technology, provided international climate policy remains important.

Future development of hydrogen fuel cells is very uncertain and depends on different drivers, including energy prices, technology development and climate policy. For CO_2 reduction calculations, this option interacts with biofuels, hybrid vehicles and fuel efficiency improvements. The development of CCS furthermore is important since it creates the possibility to produce hydrogen in a climate-friendly and cost-effective fashion.

3.4   Indicated cost of selected options in residential and services

Sectors households and services accounted for 39% of CO_2 emissions in the EU in 1990 (Joosen & Blok, 2001), when indirect emissions attributable to power consumption are included. Two major options to reduce energy consumption and CO_2 emissions are 1) insulation (wall, glazing, roof and floor) and 2) efficient heating systems. Implementation of EPBD (Energy Performance of Buildings Directive, 2002/91/EC) should be taken into account. Yet as it sets no mandatory measures, the baseline is current practice according to national policy.[35] The European Climate Change Programme (European Commission, 2003) estimates that the EPBD will achieve 220 $MtCO_2$-eq reduction, of which 35-45 $MtCO_2$-eq will be realised by 2010. Most of this reduction would be achieved at negative cost.

For both options discussed here, it should be noted that costs differ considerably among countries. In general a decreasing trend in cost from northern to southern Europe due to lower labour costs is observed (Ecofys, 2005a) and CO_2 abatement costs may likewise be significantly lower in southern Europe.

3.4.1 Insulation: Walls, roof and windows

The IPCC *Third Assessment Report* (IPCC, 2001a) reports a reduction potential of 1.2 $GtCO_2$/yr at negative cost in industrialised countries and economies in transition (EIT) for residential and 0.7 $GtCO_2$/yr for commercial buildings, both for 2010 (for 2020 the figures are 1.5 and 0.9, respectively). The reductions refer to savings due to better insulation of buildings.

Joosen & Blok (2001) report potential 2010 EU-15 savings in the household and service sector due to insulation of 130 $MtCO_2$ with costs ranging from negative values to €10/tCO_2. The assumptions regarding energy prices are unclear, however, and the discount rate at which consumers implicitly would discount benefits (avoided energy costs) is assumed to amount to 4%. The analysis as such is financial from the end-user perspective.

Ecofys (2005a) note that wall and roof insulation is particularly profitable, with payback times of less than five years. Floor and front insulation are less cost-effective, with payback times up to 15 years (which is considered to constitute net benefits for households). Earlier studies from Ecofys indicate a realisable potential in EU-15 in 2010 of 70 $MtCO_2$/yr at negative cost, relative to business as usual, and 36 Mt/yr compared to the Energy Performance of Buildings Directive. For the 10 new member states, a 14 Mt/yr potential with net benefits is estimated (Ecofys, 2005b). The follow-

---

[35] It is mandatory to provide information on energy efficiency. Filling the information gap may well provide a sufficient stimulus for certain no-regret options to be implemented.

ing table gives a more detailed overview of the CO_2 abatement costs by insulation measures.

Table 3.5 Cost assessment of insulation options in three European climatic zones

Insulation	Independent/coupled*	Cold	Moderate	Warm
External	Independent	585	9	-64
	Coupled	146	-131	-166
Cavity	Not applicable	-63	187	-208
Interior (wall)	Coupled		-159	-191
Roof	Independent	-61	-185	-222
Floor/ceiling	Independent	179	-79	-148
Windows	Independent	200	300	295
	Coupled	-151	-46	-23

* Coupled refers to implementation of the insulation measure at the time the building is renovated; independent means implementation occurred at other instance.

Source: Ecofys (2005a).

In Ecofys (2005a), energy prices for households are assumed to be €11/GJ for gas, €10/GJ for oil, and €88/MW for electricity in 2002, all increasing with an average rate of 1.5% to 2032. These are end-user price estimates,[36] and therefore the cost-effectiveness is also from the end-users point of view.

Menkveld et al. (2005) report, based on comprehensive model calculations, potential savings in households and the commercial sector of 1.9 and 2.8 $MtCO_2$ in 2020 for the Netherlands with a payback time of less than 5 years (also 2.5 Mt in transport), from the end-user perspective. In the commercial sector this is mainly electricity savings, while for households half of the savings is gas consumption (insulation, mainly glazing), most of which in existing buildings. The unused potential is caused by a lack of information and a lack of investment capital (preference for other investments).

In the residential and service sector, end-user taxes (levies and VAT) are in general an important part of the energy cost to the end-user. In a social cost-benefit analysis, these should be excluded (see also section 2.2).

---

[36] While stated explicitly, it would seem from the prices level that these are end-user prices.

Boonekamp et al (2004) show how measures in Dutch households projected for 2000-10 are very cost-effective if calculated using the end-user approach (€-253/tCO₂). From an economic efficiency perspective, i.e. adding back subsidies on the implementation of the insulation measures and removing taxes on energy, costs are relatively high: €192/tCO₂. The assumed energy prices however are rather low – €3.16/GJ for gas and €30/MWh for electricity – which reduces the direct benefits and, consequently, may overstate the 'true' economic cost.

3.4.2 Efficient heating systems

Figure 3.2 shows that gas and petrol products are the main energy sources for space heating in Europe (European Commission, 2004). Together the two sources take a 75% share of the total.

Figure 3.2 Energy sources for space heating in European households in 1996-2003

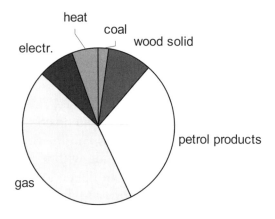

Sources: IEA (2004a) and MARKAL.

Approximately 90% of the energy consumption in the buildings sector is used for heating (and cooling) purposes, while the remainder is used for electric appliances. Just like reducing losses via insulation, improving efficiency in heat generation also deserves to receive high priority.

When natural gas is used as the heating energy source, major energy savings can be achieved by installing condensing boilers. This is based on the principle that steam produced in fuel combustion can also be used by a heat exchanger that condenses the water vapour and extracts the heat. Condensing boilers may improve efficiency considerably, e.g. from 70% to 95% (Joosen & Blok, 2001). In most countries market diffusion of condensing boilers is not very high.

This technology cannot be directly used to replace heating equipment based on oil. The additional cost of a condensing boiler is estimated to be approximately €1,000-1,600 (Ecofys, 2005a) and the emissions reduction potential in 2010 in the EU-15 would be 15 MtCO$_2$ at €50/tCO$_2$ (Joosen & Blok, 2001).

3.4.3 Barriers

Most studies report a significant reduction potential with negative cost from the residential and services sectors. A number of barriers can be identified that inhibit harnessing this potential (IPCC, 2001a):

- Fragmented market structure (many small firms, many different types of buildings, many stakeholders, division of responsibilities in building and renting);

- Building owners and renters have only limited influence on energy performance or life-cycle cost;

- No incentive to build energy-efficient, as capital costs are higher;

- Information gaps and complexity; and

- Insufficient availability of climate-friendly appliances and equipment.

Another issue is the rather low removal rate of buildings. A point estimate is that over a period of 2-3 year 175,000 dwellings (less than 0.2% of the EU stock) is removed (Joosen & Blok, 2001; 3).[37] This low demolition rate is due to the fact that policies in most European countries focus on renovation. As a result, maximum penetration levels for enhanced insulation (except glazing) are limited. Joosen & Blok estimate that the rate at which existing buildings can be refurbished (retrofit rate) is a maximum of 3% per year.

---

[37] This point estimate should not be taken as exemplary for a long time period or for a large geographical span, but is mentioned here only for illustrative purpose.

3.5   Summary

We have retained a selection of major options for subsequent cost-effectiveness analysis. Table 3.6 shows these options and summarises, where applicable, indicated CO_2 emissions reduction costs in terms of €/tCO_2 we found in the literature consulted.

Table 3.6   Overview of selected CO_2 emissions reduction options and their reduction cost as indicated by studies consulted

Option	Energy production / abatement Cost[*]	Unit	Potential EU-25	Sources
Wind	29-120	€/MWh	Large (>200 $MtCO_2$/yr) for on- and off-shore	Hoogwijk (2003), IIASA (2005), ECN/MNP (2005b)
IGCC	20-29	€/tCO_2	Not explicitly stated but can be large	IEA/OECD (2005b), Lako (2004)
Biomass co-firing	45-70	€/MWh	Large (>300 TWh/yr)	Hoogwijk (2003), ECN/MNP (2005), IIASA (2005)
Nuclear	19-44	€/MWh	Very large, but high investment cost is barrier	NEA/IEA (2005), ECN/MNP (2005a)
CHP	Depends on relative prices		Very large	Smith (2001), ECN/MNP (2005a)
CCS+PCC	30-70	€/tCO_2	Very large	IPCC (2005)
CCS industry	18-70	€/tCO_2	(Very?) large	IPCC (2005)
Biofuels	60-210	€/tCO_2	Large (>10 EJ/yr)	IEA (2004b), Wakker et al. (2005)
Hydrogen fuel cells	17	€/GJ	Large (not stated explicitly)	IIASA (2005)
Insulation	large range, mostly negative	€/tCO_2	> 400 $MtCO_2$/yr	IPCC (2001a), Ecofys (2005a/b)
Heating efficiency	-200 - 50	€/tCO_2	15 $MtCO_2$/yr (2010), larger in longer term	Joosen & Blok (2001)

[*] Negative amounts indicate net revenue (gross revenue exceeding gross cost) per tCO_2 avoided.

Table 3.6 serves as a broad overview of the cost survey in this chapter. It should be read and interpreted with due caution, as the different literature sources use different cost calculation approaches and the figures therefore can be compared only to a limited extent. The reviewed documents were not all necessarily prepared for public policy advice. The guidelines presented in section 2.6 above were followed, if at all to a limited extent only.

4. Externalities

4.1 Introduction

Financial cost-benefit analysis of projects under appraisal focuses on the short- to medium-run profitability to project financiers under projected market prices and business environment.[38] Conventional *economic cost-benefit analysis*, narrowly conceived, sets out to correct financial project analysis for 'price distortions', preventing market prices from coinciding with efficiency prices, reflecting real scarcities in the national or regional economy. Distortions of market prices from efficiency prices might be occasioned by public sector intervention such as indirect taxes, price subsidies, over(under)valued exchange rates and (other) market failures preventing labour and capital markets from functioning properly. *Social cost-benefit analysis*, in turn, *sets out to adjust conventional economic cost-benefit analysis for external effects, where clearly the wider public interest is at stake.* This chapter explores some major externalities of GHG emissions-reduction project activities.

The externalities of CO_2 mitigation projects and measures are the socio-economic side-impacts that significantly affect the socio-economic position and/or well-being of individuals other than the project financi-

---

[38] Commonly used medium-term profit adequacy indicators are the financial net present value (FNPV) and the financial internal rate of return (FIRR). A short-term profit adequacy indicator, often applied by investment analysis practitioners to risky projects in the energy and mining sectors, is the capital recovery period or pay-back period (PBP).

ers.[39] Ancillary costs and benefits of CO_2 reduction measures are externalities outside the domain of climate change policy. Typically, these externalities may take on significant proportions only several years after initiation of the climate change programmes over long periods of time. Several ancillary benefits are focal points of attention in other policy areas. It is often very difficult to estimate the value of ancillary benefits in monetary terms, although progress is being made. Especially for public health benefits, knowledge is developing although major gaps remain, such as the economic valuation of health damage. This chapter discusses distinct categories of potentially major ancillary costs and benefits.

Mitigation activities might have non-negligible macroeconomic effects, e.g. impacts on GDP, income distribution, employment, trade, technology development, etc. In section 4.2 we discuss macroeconomic income effects. Technology development impacts and employment impacts will be discussed separately in more detail in sections 4.3 and 4.4. Externalities caused by pollutant emissions, such as notably SO_2, NO_x and PM, are dealt with in section 4.5. Section 4.6 lists some environmental impacts not (yet) included in the proposed standard framework. A novel aspect presented in this chapter is the introduction of a relatively simple way to make allowance for the external cost of long-term energy supply security risk (section 4.7). Section 4.8 discusses the avoided cost of climate change through mitigation and why these benefits are not included in the proposed framework. Section 4.9 discusses some ancillary costs. The chapter ends with concluding remarks in section 4.10.

4.2 Macroeconomic income impacts

In principle, economic cost-benefit analysis can to a substantial extent address macroeconomic effects of specific projects and measures by proper efficiency price valuations of their inputs and outputs. For instance, the World Bank has sponsored the development of methodologies to account for income distribution impacts (Squire & van der Tak, 1975; Little & Mirrlees, 1975). Employment impacts can, in principle, be addressed by properly shadow pricing the cost of labour (see footnote 3). For estimating

---

[39] Social impacts may extend beyond the sum of individual 'utilities', due, among others, to ignorance of affected individuals and intrinsic environmental values, such as biodiversity.

the macroeconomic income effects of large programmes and broad policy measures, social cost-benefit analysis might be fruitfully applied in a bottom-up approach after decomposition into more specific activities and measures or mitigation options.

Social cost-benefit analysis is not readily suitable for detailed analysis of the structure of inter-industry effects from indirect inter-industry deliveries and induced consumption.[40] Moreover, net mitigation impacts might be exaggerated by the application of merely cost-benefit analysis when induced consumption results in substantial additional CO_2 emissions. For example, financial savings due to energy efficiency measures may lead to increased consumption: the so-called 'rebound effect'. To gain better insight into the nature of inter-industry effects and feedback mechanisms such as rebound effects, general equilibrium models (GEMs) can play a complementary role to social cost-benefit analysis. It is noted however that – perhaps succumbing to the magic of the esoteric sophistication of GEMs – policy-makers have tended to overstate the robustness of quantitative valuations of CO_2 reduction costs resulting from running GEMs. The reason is that GEMs require sweeping assumptions on the state of the world that are rather strong stylisations of real-world conditions. Model results tend to be very sensitive to assigned values for the parameters concerned and, therefore, tend to be far from robust. Furthermore, these models often tacitly assume immediate return to equilibrium situations following exogenous disturbances. In practice, however, the full impact of exogenous disturbances can have very long time lags. Moreover, these models do not properly take into account that tightening environmental regulation may trigger 'innovation offsets'. The latter are comprised of environmental *and* overall efficiency-raising innovations translating into external benefits that *may* (partially or more than) offset the compliance costs of environmental regulation (Porter & Van der Linde, 1995). For example, the introduction of the federal SO_2 allowance trading (SAT) programme in the US triggered

---

[40] *Indirect inter-industry deliveries* of installing 'zero emission' wind turbines for electricity generation are e.g. deliveries by a wind turbine manufacturer and O&M (operating and maintenance) service companies, who in turn need certain inputs, etc. All these activities generate value added that leads to *induced consumption*, while delivery of consumption goods again creates value added which, in turn, further raises the level of induced consumption.

SO$_2$-efficient innovations, mostly of an incremental rather than a radical nature. This scheme, leaving much flexibility to obligated parties, unleashed among the latter the use of unexpected potential creativity (business acumen), which in the initial phase rendered the cap to be achieved much more easy than foreseen. A similar phenomenon may be unfolding at present with respect to the EU emissions trading system (ETS). In the absence of banking opportunities from phase 1 into phase 2 this may well be witnessed if and when the scheme's allowance price will decline further, especially towards the end of phase 1 in 2007.[41]

4.3 Climate change policies and technological development

Can smart climate change policies stimulate technological development? The 'Porter hypothesis'

Porter & Van der Linde (1995) hypothesised that stringent but smart environmental regulation might lead to improved competitiveness of a nation.[42] In the context of this report, we might think of smart GHG-emissions-curbing policies, including the EU ETS, that according to this 'Porter hypothesis' are poised to spur GHG-saving technology. Porter & Van der Linde provided intuitively-convincing inductive arguments based on case studies, but fall short of formally proving the Porter hypothesis (see also Annex 2).

Can climate change policies and measures speed up technological learning?

Let us set out some concrete cases of technical learning in the field of electricity generation. Two different aspects of technological development (i.e. technological learning and experience) with respect to generation of renewables-based electricity (RES-E) can be discerned (Junginger, 2005):

---

[41] It cannot be totally excluded however that certain external factors, e.g. a weaker impact of the Linking Directive (less CDM credits brought into the system) than anticipated, may more than offset the impact of this phenomenon.

[42] The competitiveness of a nation at the aggregate level would mean the nation's evolution regarding 'average' productivity (the value per unit of labour and per euro of capital invested).

- specific investment cost reductions (€/kW), which might be stylised by fitting 'experience curves' with 'progress ratios' or 'learning rates' as critical parameters;[43] and

- other developments such as gains in efficiency or load factor or reduced O&M costs, which are not reflected in progress ratios, but which nonetheless may have a downward impact on the cost per unit of energy (e.g. €/kWh).

Let's assume that the stylised experience curves provide a reasonable reflection of reality indeed. Then each time the installed base of an electricity generation technology in question doubles, the cost per unit of capacity decreases by a certain fixed percentage. Typically, technological learning rates in renewable generation technology are higher than for conventional generation technology. The stability of the 'learning rate' ('progress ratio') over time is a contentious issue, but there is no doubt that economies of scale in production in tandem with innovations make for cost reductions over time. *Hence, climate change activities foster 'endogenous' technological learning in the field of renewable electricity generation. This holds true for other GHG-emissions-saving technology as well.*

An example for which technological learning is an important issue is clean coal technology. Currently IGCC plants achieve an efficiency of approximately 45%, but as research continues and higher boiler temperatures can be used, it is expected that the efficiency may increase up to 60%, increasing the economics and reducing the environmental impacts. Regarding end-of-pipe air pollution reduction equipment, capital cost reductions of 50% have been achieved in approximately 20 years (Van der Zwaan, 2005). Due to the similarity between these technologies and CO_2 capture equipment, it is expected that the costs for CCS will also decrease. A learning rate of approximately 10% can be expected for CO_2 capture systems, which means a 10% reduction in specific costs for each doubling of in-

---

[43] The progress ratio is a parameter describing the rate at which specific capital costs decrease for each doubling of installed capacity, e.g. a progress ratio of 0.8 implies a 20% cost reduction for each doubling. This translates into a learning rate of 0.2 (Junginger, 2005). Given an S-shaped market penetration curve, however, it can be expected that at more mature commercialisation stages progress ratios will taper off. Hence prudence is in order when using progress ratios for market-penetration projection exercises.

stalled capacity (Van der Zwaan, 2005). This shows the importance of substantial public support (but with clear monitoring and review milestones) for R&D and market introduction of technologies regarded as promising.

Renewable energy options for which technological learning is crucial include offshore wind power and biomass gasification. Both technologies are likely to be important in Europe if set and envisaged RES-E targets are to be achieved. Nevertheless, these technologies are commercially and even to some extent technologically still immature at present, although a large cost reduction potential exist. A modelling exercise suggests that cost reductions (€/kW installed) of 27-35% for offshore wind and 48-67% for biomass gasification are possible by 2020 when for a EU-25 wide target of 24% would apply to RES-E. Also the cost for on-shore wind may decrease substantially by 2020 to approximately €450/kW (Junginger, 2005).

Is eco-efficient, GHG-saving technological development relevant for sustainable development?

Jan Tinbergen, first Nobel laureate in economics,[44] expressed in mathematical terms the relationship between achieving sustainable development and the depletion of exhaustible resources (including energy resources) as a race of resources-saving technological development against time (Tinbergen, 1973). If the pace of technological development were to fall short of the required rate, given the population growth and desirable levels of total (world) consumption, sustainable development was not to be achieved. He published his by now almost-forgotten but still quite relevant essay in the aftermath of the dismal *Limits to Growth* study (Meadows et al., 1972). The latter blockbuster stirred both scientists and policy-makers during the era of the first and second oil crises to (temporarily) render the perceived threat posed by the depletion of exhaustible resources a hot policy issue. Tinbergen stressed the crucial role of technology development, an aspect neglected in the Meadows model. Technology development can for example have a significant impact on the extraction trajectory and the price trajectory of exhaustible resources (see Annex 3).

---

[44] An honour he shared with Ragnar Frisch.

4.4   Employment

Employment is sometimes presented as an important external benefit of implementing climate change policies. However, since it is often difficult to estimate the net employment benefits, the argument needs to be developed properly. Often the limelight is on direct employment creation of promoted activities, e.g. in the renewable energy domain. Yet to obtain the overall picture, one should also include:

- employment lost in the supply chain of energy carriers replaced,

- net indirect employment impacts in input-delivering industries and

- net secondary employment impacts in consumer industries.

Typically net indirect employment impacts are higher for e.g. renewables-based technologies compared to fossil-fuel based competing technologies (higher domestic content). However, high rates of effective protection (high subsidy rates) may partially or more than offset positive direct and indirect employment effects. Credible total employment impact studies are hard to find. We briefly discuss below two European studies that attempted to provide a genuinely overall employment impact analysis.

From a large study, encompassing 44 case studies in nine EU-15 countries, input-output analysis and general equilibrium modelling, Wade & Warren (2001) find that *energy efficiency programmes* have significant positive net employment effects. Direct employment gains were quantified to 8-14 person-years per million € of total investment. These jobs were often in groups that were prioritised in employment policy, e.g. low-skilled manual labour. They also note that employment effects are rather case-specific and diverge substantially on a per unit of investment basis.

An ECOTEC-led consortium has carried out the only truly comprehensive study so far on the employment impact in the European Union of the production and use of *renewable energy sources* (ECOTEC, 2003). In the framework of this study, an apparently appropriate input-output model method (RIOT) has been applied to assess employment and value-added impacts of RES promotion policies in the EU-15. However, the model outcomes are based on just one scenario, which provides an implausibly rosy medium-term picture. Rather optimistic assumptions made of the future trajectory of additional costs of a number of major renewable energy technologies including notably biomass technologies, constitute one major factor underlying the positive employment outcomes. Moreover, a contentious assumption explaining a large part of the projected positive employment impact is that the expansion of biofuels feedstock occurs without displacing employment in conventional agriculture. These assumptions lead to model outcomes, which appear to grossly underrate the negative indirect effects of RES stimulation. Even the positive sign of the medium-term total employment impact (661,000 full-time job equivalents for the EU-15 in 2010) does not seem to be robust, because of the great sensitivity of the outcome to assumptions such as the cost-reducing technological developments referred to above. A comprehensive study, such as ECOTEC (2003) , but with *several* (plausible) underlying scenarios, could have yielded genuinely meaningful results.

A final observation is that the potentially beneficial employment effects of unanticipated technology innovations cannot be duly captured in modelling exercises. At a minimum, this externality should be prominently mentioned as a significant qualitative consideration in the summaries of employment studies, such as the ones referred to above.

4.5   Air pollution

Particulate matter, sulphur dioxide and tropospheric ozone are the health-affecting substances most studied in epidemiological studies. It is however likely that other substances, such as lead, mercury and other metals, have an impact on human health as well (Ezzati et al., 2004). Carbon monoxide is also important in urban environments. NO_x is important as a precursor for small particulates and ozone, and as one of the major substances responsible for acidification and eutrophication. The WHO (2000) estimates that 1,200 life years are lost per million capita (urban population) in the EU-25 due to urban smog, only on account of particulate matter (PM). Bouwman & Van Vuuren (1999) estimate that over 35% of ecosystems exceeded in Europe the critical load[45] of acidification in 1992. Although acidifying emissions are decreasing, the magnitude of their negative impact will not have been much ameliorated over the last few years.

For making a social value assessment of air pollution reduction co-benefits of climate change policy, two main approaches can be pursued. The *damages cost approach* assesses the impact of air pollution (e.g. on health) and gives an economic value to these damages. Another approach

---

[45] Defined as the maximum pollution load at which the exposed system is not damaged.

uses *avoided abatement cost* for achieving baseline air pollutant standards. According to welfare economics, it is not optimal to internalise all damages costs in the pricing system. The socially optimal point up to where the damages costs should be internalised coincides with the point where the marginal abatement costs of air pollution reduction are equal to marginal benefits. These benefits reflect the social value of the adverse impacts of the marginal (last) unit of air pollution. In practice, due to quantification problems, is difficult to determine the optimal level of air pollution and the corresponding optimal marginal abatement cost level.

In general, combustion of fossil fuels has significant external costs due to air pollution causing health and ecosystem impacts. In contrast, renewable energies exhibit very small such externalities. The nuclear fuel cycle also features small external costs, although it is not so clear whether all the significant impacts of this fuel cycle have been duly quantified. (ExternE, 2005).

Rabl & Spadaro (2000), as part of the ExternE project, give estimates for damage factors for air pollutants, as shown in Table 4.1, below applying the *damages cost approach*. The damage costs mainly result from impacts on morbidity and respiratory diseases, which are translated into 'years of life lost' (YOLL) by epidemiological studies. The authors then make an economic valuation of the YOLL multiplying with the 'value of statistical life' (VSL) based on the willingness to pay (in Europe). The value of 1 YOLL is estimated to be €83,000, calculated from VSL of €3.1 million. A discount factor of 3% is used. Even without considering climate change, ExternE indicates very significant adverse impacts of current patterns of power generation for human health. As for climate change damages, ExternE puts the marginal damages cost per tCO_2-eq. if the Kyoto targets are met at €29/tCO_2-eq.

For several conventional generation technologies, the ExternE project has elaborated on the damages cost approach to provide estimates of total damages cost per unit of generated electricity. For certain generation technologies, such costs may well be of the same order of magnitude as, or even higher than, the unit generation cost excluding these externalities (see Table 4.1). Note that the figures arrived at by ExternE are estimates of *total* per unit cost of air-pollution related damages and *may therefore overstate the socially optimal* per unit cost of these damages.

Table 4.1   Damages cost factors according to Rabl & Spadaro (2000)

Pollutant	Damages cost €/kg pollutant	PCC plant with end-of-pipe abatement €/MWh	Gas combined cycle, low-NO$_x$ burner €/MWh
Particles (PM)	15.4	3.1	0
SO$_2$	10.2	10.2	0
NO$_x$	16.0	32.1	11.2
CO$_2$-eq GHG	0.029	27.3	12.5
Total		73	24

The ExternE valuation of the total marginal damage costs is obviously highly dependent on the VSL assumptions and the discount rate. Moreover, it is location-specific and has inherent subjective judgements. Furthermore, it is stressed that ExternE has made an (as such credible) attempt at quantifying total marginal damage costs. As a reference, Table 4.1 presents ExternE damages estimates both in tonnes of air pollutants and in euros. In our numerical example given later in chapter 6, we take the ExternE estimate to be the upper limit of those social cost of air pollution externalities that should be internalised in social cost-benefit analyses.

Vito (2004) estimates external cost of NO$_x$ and SO$_2$ emissions due to acidification and eutrophication, based on willingness to pay for protection of land. It appears that external cost estimates are approximately 10% higher for NO$_x$ and 4% for SO$_2$ on average throughout the EU-15 as compared to the external cost as assessed by ExternE.

Van Vuuren et al. (2006) explore the value of ancillary benefits of Kyoto Protocol implementation in Europe using the IMAGE and the RAINS models for an integrated assessment. Using the avoided abatement cost approach, they estimate that the costs of air pollution control (SO$_2$, NO$_x$, VOC and PM$_{10}$) in 2010 add up to approximately €89 billion. In doing so, they assume that the EU National Emission Ceilings Directive[46] and the Gothenburg Protocol targets will be met. Western European countries bear 80% of this cost; 57% of these abatement costs are associated with mobile sources. Van Vuuren et al. (2006) argue that about half the GHG abatement

---

[46] Under this Directive, member states have mandatory emission targets for NO$_x$, SO$_2$ and NH$_3$ (acidifying emissions) for 2010.

costs can be offset due to reduction in air pollution. The cost of Kyoto Protocol compliance is highest if carried out only with domestic measures (€12 billion), but the benefits for reduction in air pollution (in Europe) are also largest (€7 billion, i.e. approximately 8% of the air pollution target AP compliance cost).

The International Institute of Applied Systems Analysis (IIASA) is a leading organisation on air pollution modelling. The national emissions ceilings in the EU member states are based on IIASA's RAINS model. IIASA uses comprehensive marginal abatement cost curves to estimate compliance cost and which technologies are likely to be used in meeting the target ceilings. In recent years, IIASA researchers are developing GAINS, a model that can calculate synergies and trade-offs between technologies to reduce air pollution and GHG emissions, using the avoided abatement cost approach. Cost data used in our numerical example (chapter 5) are partly based on i) IIASA's technology cost data and ii) IIASA's figures on interaction between air pollution and climate change mitigation.

4.6 Other (environmental) benefits and costs

Significant other environmental benefits resulting from measures to reduce greenhouse gases can be identified, but these benefits are rather hard to capture in a quantitative valuation framework. According to IPCC (2001b), implementing GHG mitigation policies in the transportation sector may well reduce urban congestion. Consequently, urban congestion reduction may constitute a significant ancillary benefit. In addition, reduction in traffic accidents (mortality and morbidity) may be a significant positive externality arising from GHG policies targeting the transport sector, especially those policies focusing on enhancing public transport.

Sustainable forest or agricultural land management – besides acting as carbon sinks – is beneficial for the local environment by enhancing water supply, combating soil erosion and protecting habitats.

Table 4.2 gives a broad categorisation of other (potentially significant) environmental benefits.

According to IPCC (2001b), implementing GHG mitigation policies in the transportation sector may well reduce urban congestion. Consequently, urban congestion reduction may constitute a significant ancillary benefit. In addition, reduction in traffic accidents (mortality and morbidity) may be a significant positive externality arising from GHG policies targeting the transport sector, especially those policies focusing on enhancing public transport.

Sustainable forest or agricultural land management – besides acting as carbon sinks – is beneficial for the local environment by enhancing water supply, combating soil erosion and protecting habitats.

Table 4.2   Ancillary costs/benefits to be qualitatively assessed

Benefit/cost category	Example of reduction option
Natural resources, such as forests and water	Sustainable forestry
Biodiversity	Sustainable forestry, agricultural methane policy
Waste generation	Reduction in fly-ash generation by coal-based power
Urban congestion and noise reduction	Traffic management
Visual impact	Wind, fossil or nuclear power plant
Risk (e.g. accidents)	Electricity demand reduction, as compared to new nuclear power plant
'Technological learning': benefits for more opportunities to implement environmental-friendly technologies within and outside Europe	In particular all long-term options, such as PV and hydrogen fuel cells
Comfort of living	Insulation

4.7   Energy supply security

Energy security of supply can be defined as "the availability of energy at all times in various forms, in sufficient quantities, and at affordable prices" (IEA, 2005). This concept can refer to the prevention and mitigation of short-run emergencies as well as the reduction of long-run energy supply security risk:

- Prevention – and introduction of adequate impact mitigation procedures in the event of – of immediate supply emergencies (huge recent power black-outs in the US and Italy; sudden malfunctioning of natural gas supply through major pipelines from Russia or Algeria to the

EU, for example because of terrorist acts or use of the embargo weapon in political conflicts).

- Prevention of over-exposure to long-term energy supply security risks as reflected by 1) a strong structural upward trend in weighted overall energy prices to end users and 2) high energy price volatilities in the EU out of sync with major overseas competitors.

Long-term supply security can be improved through a multifaceted approach, including the use of the following options (Jansen et al., 2004; IEA, 2004b):

i) *Diversification of energy sources.* Special attention to limit over-dependence on oil and natural gas and to stimulate promising re-newable technologies and distributed generation in ways that are consistent with dynamic economic efficiency. Nuclear energy is a contentious option with idiosyncratic problems, but discarding this option altogether raises the urgency of the long-term energy supply security issue. 'Clean coal' technology in step with 'carbon capture and storage' is further discussed below.

ii) *Diversification of oil and gas sourcing* by mode and suppliers (pipeline, ship haulage). Special attention to limit over-dependence on suppli-ers in countries with unstable political regimes and to limit depend-ence on vulnerable transport trunk routes and hubs.

iii) *Improving demand response opportunities* through 1) well-functioning spot markets with an evolution from national markets towards supra-national regional markets and ultimately EU-wide markets and 2) in-novations driven by increased interconnectivity between end users, traders and national or EU-based suppliers to be enabled by IT infra-structure.

Long-term energy supply security risks can be considered to be ap-preciably greater than the overall picture presented by leading official en-ergy policy information agencies such as IEA and EIA. These agencies ap-pear to exaggerate the possibilities for increasing the world's proven re-serves and output levels of oil. Information from the US Geological Survey (USGS), on which these agencies rely importantly, suggests much higher ultimately recoverable reserves than most other sources. Furthermore, the average additions to proven oil reserves per 'wild cat' appear to have de-creased substantially over recent years. Western oil companies need to strongly increase their upstream expenditure to maintain access to satisfactory reserve levels. Moreover, the percentage of total proven reserves under control of western oil companies is decreasing, while the role of state oil companies in politically unstable countries is gaining weight. The 'proved resources' statistics of Middle East oil and gas producers are to a large extent not verified by reliable third parties and may have been exaggerated for political purposes (e.g. acquiring more political leverage, obtaining a higher OPEC quota, etc.). Oil demand to meet requirements of low-duty vehicles in China, India and other developing countries is growing rapidly, while the prospects for worldwide penetration of alternative motor fuel in the next two decades are bleak. Given the oligopolistic nature of the world oil market, all the ingredients are present to suggest highly volatile world oil prices with a structurally upward tendency. The global natural gas resource base is characterised by a slightly less uneven distribution and a slightly stronger resource base compared to current demand levels. Still, the prospects for the natural gas market in the EU are similar: high volatility with a long-term trend of firming real prices. [47]

A report commissioned by the WWF projects that *a stringent climate change policy in the EU has substantial energy supply security co-benefits for the EU*. Reduction of the demand for oil in a scenario with 33% GHG reduction compared to 1990 as compared to the baseline scenario is projected to slash the EU oil import bill by $60-120 billion by the year 2020 and will reduce appreciably the dependency of the EU on imports of oil and natural gas from politically less stable countries (Wuppertal Institute, 2005).

We concur with the WWF and the Wuppertal Institute on the significance of energy supply security co-benefits of European climate change policy in the medium and long term. Exhaustible resources with global production, such as oil and natural gas, are expected to peak in the foreseeable future (in 10-30 years from now for oil and 40-60 years from now for natural gas). Moreover, these resources are distributed quite unevenly geographically. This distribution makes for an oligopolistic market with strong market power exerted by suppliers from politically (potentially) instable countries. This not only makes for serious energy supply security risks; it can be argued that it also results in related external costs in terms of public expenditures in military stabilisation operations in the Middle East and

---

[47] See section 4.2 for some theoretical arguments pointing in the direction of price trajectories for oil and natural gas that will notably rise in the longer term.

neighbouring countries. Ogden et al. (2004) argue that military expenses to safeguard access to Middle East oil can be used to make a conservative estimate of energy security external cost. They come to $15-$44 per barrel for the US, based on $20-60 billion expenses annually (based on references from the year 2000, which does not include recent military activities) and a share of 20% of global imports for the US.

In the context of the ExternE programme, Markandya & Hunt (2004) aim to quantify economic externalities of security of supply. They conclude there is "some correlation between a higher oil price and lower GDP growth rates with a one to two year lag". Also they argue that high price volatility acts as a disincentive for investments in the oil industry. Accurate estimation of external costs, however, remains extremely challenging.

Therefore the case for making allowances for the external cost of long-term ESS risk in social cost-benefit analyses in a European context would seem a compelling one. Yet, market prices reflect all relevant information, at least all public-domain information. On the other hand, market parties tend to give more weigh to short-term aspects than to long-term aspects. Short-term aspects in the oil and gas market are interrelated boom-bust upstream and downstream investment cycles on the one hand and strongly bullish and bearish price expectations on the other. Also the fairly inelastic demand for oil and natural gas is a major underlying factor. The boom-bust investment cycles relate to the major indivisibilities in typically giant resource development and transportation infrastructure projects. With the possible exception of oil and gas prices during periods when markets anticipate or face *short-term* supply constraints, we would argue that *long-term* ESS risks from a (European) societal point of view are not adequately factored in prevailing gas and oil prices. The major arguments in favour of this position include the following:

1.   The EU's dependence on a dozen or less external oil- and gas-supplying countries is already high and will rise further. The same trend is unfolding for other major importers, e.g. the US, China, Japan and India. Consequently, the already-significant ability of the world's key oil and gas suppliers to exert market power is due to increase further.

2.   The huge transfers of windfall rent income from oil- and gas-consuming to producing countries *may* lead to political instability on the receiving end, transmitted abroad. Moreover, sudden rebalancing actions by major oil-producing countries in the investment portfolio

of their foreign reserves could upset overseas financial markets. For instance, what would happen to the US and other industrialised-countries' economies if and when financial investments in liquid US dollar-denominated assets by oil-producing countries would be dumped within a short span of time?

3.   In the key exporting countries, mostly state-monopolist corporations under close tutelage of the central governments concerned are put in charge of extraction and trading of national oil and gas resources. Yet politicians *may* make commercial decisions that are rational from their political perspective but suboptimal from an economic rationality perspective.

4.   Great uncertainty exists about the actual rate of depletion of ultimately recoverable oil and gas reserves throughout the world.

We propose that a certain base year ESS risk premium to oil and gas use be set at a level reflecting preferences of EU and member state policy-makers. For expository purposes we assume the rates shown in the Table for the base year of the numerical example in Chapter 5, i.e. year 2003. As these values are inherently subjective, in practice they can be set in a dialogue with policy-makers and other stakeholders. Note that the premium increases by the central social discount rate, as explained below.

The assumed social ESS risk premium costs for oil have been set higher than natural gas considering:

i)   the currently higher depletion rate of proven reserves of oil relative to natural gas and

ii)   the slightly less concentrated distribution of proven natural gas reserves worldwide.

As for the time trajectory of the ESS risk premiums, we propose:

i)   The premiums rise at a compound pace, using the social discount rate. This follows the Hotelling rule regarding the future value trajectory of exhaustible mineral resources.

ii)   The projected fuel price upper-bounds are presumed to fully reflect the social cost of ESS risk. Hence, the ESS risk premium used is subject to the constraint that simulated fuel price plus SSE premium cannot exceed the projected fuel price upper-bounds.

These rules ensure that except for periods of projected 'high' fuel prices, oil and natural gas use will be penalised in the social cost-benefit analyses by the pre-set ESS risk premium. The graphs in Annex 5 show the trajectories of the oil and gas prices including the risk premium.

For reasons of resource constraints in this limited study, we have only applied ESS risk premiums for the highest-priority exhaustible fossil fuels. ESS risks with regard to uranium and, even more so, for coal are much less pronounced than is the case with oil and natural gas. Reserve-production ratios with regard to uranium and coal are, conservatively estimated, over 100 years and 250 years, respectively. Moreover, reserves are less concentrated in countries with seemingly unstable political regimes. Yet the proposed ESS risk valuation methodology can be applied equally well to e.g. coal and uranium resources.

4.8 Avoidance of climate change cost

Avoiding dangerous climate change is the primary objective of climate change policy. The European Union has repeatedly reiterated its aspiration to take the lead in shaping such a global climate change policy regime.

In addressing the issue of adaptation, IPCC (2001b) mentions the following on impacts of climate change:[48]

Projected adverse impacts based on models and other studies include:

- A general reduction in potential crop yields in most tropical and sub-tropical regions for most projected increases in temperature

- A general reduction, with some variation, in potential crop yields in most regions in mid-latitudes for increases in annual-average temperature of more than a few °C

- Decreased water availability for populations in many water-scarce regions, particularly in the sub-tropics

- An increase in the number of people exposed to vector-borne (e.g., malaria) and water-borne diseases (e.g., cholera), and an increase in heat stress mortality

---

[48] Note that in addition to these mostly negative impacts, regional positive impacts may also occur, such as increased average agricultural yields across Europe if the temperature rise is below 2°C (Watkiss et al., 2005).

- A widespread increase in the risk of flooding for many human settlements (tens of millions of inhabitants in settlements studied) from both increased heavy precipitation events and sea-level rise

- Increased energy demand for space cooling due to higher summer temperatures.

Ogden et al. (2004) use GHG damage cost based on least-cost reduction options that achieve deep cuts, estimated at \$66-170/tC and a mean of \$120/tC (33 €/tCO₂). These values should be considered as projected abatement costs, not as projected damage costs. As indicated by Azar (2003), damage-cost calculations are surrounded by very large uncertainties and inherent subjective judgements.

It is emphasised that the valuation of benefits of avoided climate change (or 'cost' of climate change) invariably involves large uncertainties and subjectivity due to:

- Choice of the discount rate (bias towards overvaluation of near-term cost and benefit cash-flows; risks far in future tend to be undervaluated; possibly a parallel can be drawn with conventional nuclear waste valuation);

- Any monetary value attached to human life, including differences between western (rich) and developing (poor) countries;

- Difficulty of assigning value to the loss of biodiversity/ecosystems?;

- Huge uncertainty regarding the occurrence of events such as cyclones, sea-level rises and droughts; and

- Huge uncertainty exists regarding the impacts of such events and subsequent impact valuation.

Watkiss et al. (2005) note that uncertainties regarding climate change impact assessments have two important dimensions: uncertainty in predicting (i) the physical effects and (ii) the economic valuation of the physical effects. They conclude that most studies have an incomplete impact coverage in both dimensions and hence underestimate the net social cost of climate change impacts. Based on Tol (2005), who carried out a meta-assessment across 28 cost studies, Watkiss et al. (2005) project a mean climate change damage cost of €25/tCO₂ and a 95-percentile damage cost of €96/tCO₂-eq. They also note that marginal cost of GHG emissions are likely to increase by 2-3% annually, which can be explained by the likeliness that impacts increase with rising emissions levels.

On the other hand, also the magnitude and uncertainty of climate change impact costs should be put in perspective. As concluded by Azar and Schneider (2002), the GDP loss attributed to climate change mitigation would be 3-6% in 2100 for 75-90% GHG reduction. This would delay income growth merely by a couple of years, so that according to Azar and Schneider we would be ten times richer in the year 2102 instead of 2100.

In addition to the debate about monetary valuation of avoided climate change, a social dilemma exists regarding the asymmetry in the incidence of abatement cost and abatement benefits. The direct cost of reducing greenhouse gas emissions is borne by a national or regional economy deciding to finance climate change mitigation policy measures, while the benefits are global and accrue to future generations.

So far the proposed standard CBA framework does not include the externality of avoided climate change costs. This externality is likely to be quite significant but at the same time it is highly uncertain. For purposes of practical policy design, the (still limited) readiness to put aside public money for climate change action programmes, policies and measures is taken as a reflection of a (lack of a) sense of urgency of the climate change issue. Awareness raising activities should narrow the gap between broadly held perceptions and latest scientific insights. Given these presumptions, all social cost to be incurred should ideally be accounted for net of those for addressing climate change impacts. Inclusion of (avoidance of) high externality costs regarding climate change impacts risks to arise the suspicion among climate change sceptics that the proposed framework is flawed. Conversely, a focus "merely" on the (significant) co-benefits may well widen the acceptance of the proposed framework. It is noted that leaving out the climate change externality is done for practical policy design rather than for fundamental reasons: the framework can facilitate its inclusion if and when deemed appropriate.

4.9 Ancillary costs

The public costs of running and monitoring climate change programmes can be quite substantial. They include the staff and material cost of the climate change unit, expenditure on demonstration projects, awareness raising activities, the preparation, implementation, and enforcement of GHG-reduction-related standards, (to some extent) energy efficiency audit programmes, GHG-emission labelling programmes, etc. To the extent that these costs are not attributed to the specific activities and measures encompassed by the climate change programmes in question, they can be deemed to be ancillary costs. Typically, in cost estimates of GHG mitigation activities the public climate change programme cost are highly underrated if taken into consideration ate all.

Another type of ancillary costs relate to projects proposed to become eligible for the project-based flexible Kyoto Protocol instruments, CDM and JI. Both project investors and public authorities involved in credit certification procedures have to sustain substantial dead weight (public) regulatory and (private) public relations cost. These external costs per tCO$_2$ abated relative to the approved baseline are difficult to estimate but high. Yet they appear to show a decreasing tendency. Investors become more aware of the "red tape" public relations costs involved and tend to internalise procedural costs they have to sustain increasingly well. Furthermore, procedural CDM and JI certification costs might well go down by streamlining certification procedures as a result of the Conference of the Parties held in Montreal in December 2005 (COP-11). Moreover, emergence of specialised service providers and aggregators of small-scale CDM projects help to reduce transaction costs.[49]

4.10 Summary

A wide range of distinct externalities has been reviewed on their significance and suitability for inclusion into the proposed analysis framework. These include:

- Macroeconomic impacts of GHG emissions. Proper cost-benefit analysis can largely account for these impacts.

- An exception is formed by technological development and innovations. Stringent GHG reduction policies can importantly stimulate technology development and innovation that reduces demand for fossil fuels. A high sense of urgency would seem to be in order to design and implement proper policy frameworks that foster acceleration of exhaustible-resource-saving innovations. So far, our numerical example of the proposed standard framework (set out in the next chapter) accounts for the technology dynamics through technical

---

[49] See for further information on the CDM, Egenhofer et al. (2005).

learning in an exogenous way. To endogenise technical learning is clearly beyond the scope of this study.

- Results of a literature scan suggest that the overall employment impact of energy efficiency improvement programmes is positive. In the medium term the overall employment impact of renewable energy stimulation programmes appears to be ambiguous, however. Nonetheless, net employment benefits are expected at longer time scales, as many renewables-based technologies experience much faster cost-reducing technological progress than competing non-renewable technology. Existing studies on the effect of climate change policies on employment can, however, be criticised on many grounds, and their results should be interpreted with utmost care.

- GHG emissions reduction policies and measures have significant benefits for air quality, as pointed out by a large number of literature sources. Including reduced abatement cost for air pollution reduction may offset GHG mitigation cost for a substantial part.

- GHG reduction policies and measures have significant benefits in terms of improved long-term energy supply security.

- The public costs of running climate change programmes and certification of GHG emissions reductions should be accounted for when specific climate change mitigation policies are considered.

Generally, it is very difficult to attach credible monetary values to the aforementioned effects. Yet key decisions on the design of climate change programmes are often taken on the basis of key summary figures, such as cost per tonne of CO_2 reduced. Therefore, it should be seriously attempted to at least include credible minimum estimated monetary values for major externalities to the extent possible. In this chapter it was explained how the external (negative) costs of air pollution impacts and energy supply security impacts have been internalised in the numerical example set out in the next chapter.

The proposed standard CBA framework does not include the externality of avoided climate change costs. This externality is likely to be quite significant but at the same time it is highly uncertain. For purposes of practical policy design, the (still limited) readiness to put aside public money for climate change action programmes, policies and measures is taken as a reflection of a lack of a sense of urgency of the climate change issue. Awareness raising activities should narrow the gap between broadly held perceptions and latest scientific insights. Given these presumptions, all social cost to be incurred should ideally be accounted for net of those for addressing climate change impacts. It is noted though that leaving out the climate change externality is done for practical policy design rather than for fundamental reasons.

5. Application of the proposed methodology

5.1 Introduction

This chapter presents results of "integrated" cost-benefit analysis of selected CO_2 reduction options from a social perspective. Conventional cost estimates of CO_2 reduction options do not include estimates of external effects set out in the previous chapter. At best, qualitative statements are made that no allowance has been made for such effects in monetary terms. The social cost-effectiveness analysis set out hereafter is carried out in two stages. In the first stage, conventional analysis is performed of the incremental cost of selected CO_2 reduction options per tonne of CO_2 avoided, compared to a specified, typical reference option. In the second stage, allowance is made for major external effects, that is: effects regarding

- air pollution;
- depletion of exhaustible fuel resources sourced primarily in politically unstable regions;
- technical progress with respect to CO_2 reduction options and reference options.

The analysis in this chapter provides evidence in support of the view that the inclusion of major external effects in quantitative net cost estimates is of great importance for the appraisal of the costs and the cost-effectiveness of climate change programmes to society at large. The options considered in this chapter relate to (i) electricity generated in the power and industry sectors, (ii) automotive fuel and (iii) the buildings sector.

As for the use of hydrogen as automotive fuel, the cost estimates on this option are surrounded by extremely high uncertainty on prospective technological and infrastructural developments. This option is therefore not further elaborated on in this chapter.

In an attempt to adequately reflect uncertainty in the cost calculations, we have used three values, which are derived from uncertainty analysis using @RISK software for each assumption in the calculations. The abatement cost calculated is then expressed in a 2.5-percentile, mean, and 97.5-percentile value. Rather large uncertainty ranges are the result of this approach.

In this chapter, first the abatement cost calculation methodology is explained, after which the general and technology-specific assumptions are discussed. In the next sections, cost estimates excluding (5.4 and 5.5) and including externalities (5.6 and 5.7) are given. Figure 5.1 shows how the results will be presented: the 'storyline' of our study.

Figure 5.1 Step-wise setup of numerical analysis framework and corresponding chapter setup

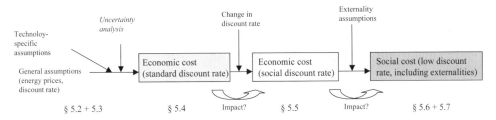

5.2   Abatement cost calculation methodology

In the numerical application of the proposed approach for determining social abatement cost for GHG reduction (see Section 2.6), the methodology for our calculations consists of:

- Input variables, including uncertainty estimates reflected in triangular distributions;

- CO_2 abatement cost calculations, based on 1) Net Present Value calculations as commonly applied in economic analysis to calculate electricity production cost of each power option and 2) biofuel production cost;

- Monte Carlo simulation based on @RISK software to translate the uncertainty in inputs into uncertainty in the abatement cost outcomes;

- Assumptions to quantify externalities, namely air pollution and energy supply security, for each option. These estimates are added to the (economic) CO_2 abatement cost

This methodology is used for those options for which it was possible in the context of this limited study: electricity sector options and biofuel. For the CCS options, insulation, and heating efficiency it was not possible to carry out detailed calculations due to a lack of reliable data; instead different estimates from the literature were taken to reflect uncertainty in the abatement cost (see further Section 5.4).

5.2.1 Input variables

Table 5.1 gives a brief overview of the variables on which our cost estimates for the electricity sector are based. In Section 5.3 the assumptions are explained further.

Table 5.1 Explanation of variables underlying cost estimates of electricity generation options

Variable	Explanation
Discount rate	Different value ranges are taken to reflect a standard or socio-economic analysis
Specific investment cost	Up-front investment per kW_e capacity in 2010
Investment-learning	Projected decrease in investment cost per year (see 5.3.4)
O&M variable	Variable operating and maintenance cost per MWh
O&M fixed	Annual operating and maintenance cost per kW_e installed capacity
Capacity factor	Assumed hours per year availability
Efficiency	Conversion efficiency (primary to electrical energy)
Efficiency-learning	Projected increase in efficiency (in %-point/yr), see 5.3.4
Fuel	Cost of fuel input (primary energy equivalent) in $€/GJ_p$
Lifetime	Estimated average plant lifetime
CEF	Carbon emission factor

For each of the variables (except plant capacity) a triangular uncertainty distribution is defined based on the different values mentioned in the literature. The range of variable values found in the literature is translated into a triangular distribution by taking the lowest estimated value as the 2.5% value and the highest as the 97.5% value. The mean is the average of estimates in the literature. We emphasise that this procedure does not yield a reliable distribution for all the variables; for this end many more literature sources would have been needed. However, for the purpose of this exercise, which is to demonstrate the proposed abatement cost calculation methodology, the distributions give a good basis for estimating uncertainty in the outcomes.

In order to estimate the fuel cost variable, fuel price trajectories are used, which is further explained in Section 5.3.2.

For the externalities of air pollution and energy supply security, the assumptions are explained in Section 5.3.6 and 5.3.7. However, the externalities are not included in the NPV calculations: after the economic cost calculations have been made, they are added to the cost outcomes as explained in section 5.2.4.

5.2.2 CO_2 abatement cost calculations

For the electricity sector, the CO_2 abatement costs are determined using the production cost of electricity ($ElecCost_{option}$) for the abatement and reference options, respectively. These are calculated using standard Net Present Value calculations, using the variables mentioned in table 5.1, and the assumed triangular distribution for each variable. It is, however, noteworthy that two different discount rate distributions (with different means and standard deviations) are used in two different sets of calculations (see Section 5.5), namely when the costs are calculated using "standard" or financial discount rates and when they are calculated using social discount rates.

For the electricity production cost of CHP, the approach used in NEA/IEA (2005) is taken. This implies that the cost of electricity is calculated by deducting the value of heat (calculated from the cost from separate generation by gas) from the total cost, while the remainder is the production cost of electricity.

To calculate the CO_2 abatement cost of an option when it is implemented instead of its reference option, the difference in (electricity or biofuel) production cost is divided by the CO_2 reduction per unit of production. For instance, the abatement cost for wind on-shore compared to PCC is:

$$Abatement\ Cost = \frac{ElecCost_{wind} - ElecCost_{PCC}}{CEF_{PCC} - CEF_{wind}} [€/tCO_2]$$

where CEF is the CO_2 emission factor in tCO_2/MWh.

For biofuels, the abatement cost calculation is based on the difference in fuel price between the biofuel and its fossil fuel reference.

5.2.3 Uncertainty in the abatement cost calculations

In order to take account of the uncertainty in input variables in the abatement cost calculations for electricity and biofuel, the @RISK tool is used (see Annex I for more information on @RISK). This programme uses simulations comparable to Monte Carlo simulations to translate the distributions given for the input variables into the output, in this case the CO_2 abatement cost. In the simulations, all input variables are varied according to their defined distribution. Each simulation yields one abatement cost outcome, and the number of simulations is such that a representative selection of each input variable has been used (the complete 'triangle' of each variable is covered). We have used 1000 simulations. Annex A.1 shows an example of a triangular distribution of an input variable.

The simulations yield a bell-shaped outcome distribution for the abatement cost. The range between the 2.5% and 97.5% values represents the 95% confidence interval of the CO_2 abatement cost. In the abatement cost results tables, we show the 2.5% and 97.5% values as well as the mean value. The 'mean' abatement cost value is calculated by using all the mean values of the assumptions. Annex A.1 contains an example of an abatement cost distribution.

5.2.4 Including externalities

To include the externalities air pollution and energy supply security in the CO_2 abatement cost calculations we used the following steps:

- Use the economic cost calculations, with the abatement cost values (2.5%, mean, 97.5%) for each option
- Calculate air pollution and energy supply security externalities for each option in monetary terms per tonne of CO_2 reduced, based on three different sets of externality assumptions (conservative, central, and maximum), as explained in Section 5.3.6 and 5.3.7.

- In case the externalities reduce the abatement cost (i.e., in the case of positive externalities from the "abatement" option relative to the "reference" option), the 'maximum' externality values are added to the 2.5% abatement cost values, 'central' to 'mean', and 'conservative' to the 97.5% value, to obtain the 'low', 'mean' and 'high' value for the social abatement cost (i.e. economic cost including externalities). In some cases, the externalities *increase* the abatement cost (i.e., we have negative externalities from the "abatement" option relative to the "reference" option). In these cases, the 'conservative' externality values are added to the 2.5% economic abatement cost value and the 'maximum' tot the 97.5% value.[50]

Ideally these three steps would be integrated into one step, including the uncertainty calculations using @RISK. However for this study this exercise would be too complex to carry out.

The terms 'low', 'mean' and 'high' are used for the social abatement cost values rather than percentages (2.5% and 97.5%) because 1) the assumptions are based on judgement in order to highlight the possible range in assumptions, in particular of ESS, and 2) the externalities are not included in the uncertainty simulations with @RISK and therefore no meaningful statement can be made regarding the uncertainty distribution. The low, mean and high value therefore should be read as indicative for the uncertainty in social cost based on the assumptions made.

5.3 General assumptions

5.3.1 Exchange rates

We have assumed a constant dollar/euro exchange rate of $US = € 0.83 in the calculations.

---

[50] By using this approach the uncertainty interval increases due to the inclusion of the externalities. It would be incorrect if the uncertainty would decrease by including externalities.

5.3.2 Discount rate

In our first calculation, we use discount rates that are commonly applied in economic cost-benefit analysis: ranging from 5 to 10% (according to a triangular distribution). In the second step we apply 3 - 5%, to reflect the social perspective. The difference in the cost outcomes shows the impact of the discount rate (see Section 5.4). As the risk free interest rate in EU capital markets tends to be of the order of 1-3 % (excluding inflation), even the social discount rates therefore include a risk margin for business risks from a social point of view. Yet it would seem inappropriate to also include financing risk and risk of short-term price oscillations around long-term structural price trajectories in social discount rates as is often tacitly done in many existing studies on climate change mitigation costs.

5.3.3 Energy prices

As mentioned in Section 2.5, future energy prices are an important uncertainty in a cost-benefit analysis. Based on Table 2.1, we designed three energy price scenarios for oil, gas, coal and uranium. We used average energy prices (for Europe, where applicable) for the period 2000-2005 as the basis, taken from CBS (2005), in €2003.

The price estimates take into consideration:

- the tendency to underestimate future oil (and natural gas) prices by international bodies;
- theoretical considerations regarding optimal extraction and consequent price setting behaviour of owners of exhaustible resources (see section 2.5.1);

Therefore, we use the latest "high price" scenario of IEA for baseline purposes. In order to construct three price scenarios up to 2030, we use price escalators (in percent increase per year), as shown in Table 5.2. The low (2.5%) and mean scenarios correspond roughly to the reference and alternative scenarios from major energy studies respectively (see Table 2.1). As price scenarios in these publications are generally conservative, we construct a scenario with higher but still realistic prices (97.5%).

Table 5.2 Energy scenario assumptions in calculations (2003 real prices)

			2.5%	mean	97.5%
2003	oil price	$/bll		33.4	
2003	gas price	€/GJ		3.3	
2003	coal price	€/GJ		1.5	
2003	uranium price	€/GJ		0.4	
	oil price escalator	%/yr	1%	2%	4%
	gas price escalator	%/yr	1%	2%	4%
	coal price escalator	%/yr	0%	0.4%	1%
	uranium price esc	%/yr	0.5%	1%	2%
2030	oil price	$/bll	44	62	96
2030	oil price	€/GJ	6.3	9.0	14
2030	gas price	€/GJ	4.3	6.2	9.5
2030	coal price	€/GJ	1.5	1.7	2.0
2030	uranium price	€/GJ	0.5	0.5	0.7

5.3.4   *Technological development*

As discussed in section 4.3, technological development and the resulting cost reductions can be a factor of great importance. This is particularly valid for newer technologies such as wind power, IGCC and CCS, where specific investment costs are projected to decrease and the efficiency likely to increase.

In this respect, it may be important to take into account in which year the technology is implemented. For example, current abatement cost for wind power may be higher compared to the cost in, say, 20 years. In order to gain insight in these effects, we carried out calculations with 2010 as the year of implementation as the base case, and compared the result with a case where 2020 is the starting year.

In our calculation, we used estimated learning rates for:

- Decrease in investment cost (%/yr)
- Increase in conversion efficiency (%-point/yr)

In Annex A.4 the difference in values for investment cost and efficiency in 2010 and 2020 are shown. These are derived from a historical progression in values and projections for further cost decrease and efficiency increase. For fossil power generation, they are derived from Lako (2004), and for wind power from Junginger (2005) and CPB/ECN (2005). Operating and maintenance cost are not likely to change significantly.

5.3.5 Technology-specific assumptions

Cost calculations were carried out for the GHG reduction options in the power sector and transport (biofuels). We did not carry out new cost calculations for the remaining options discussed in chapter 4. In these cases values derived from the literature are given in the tables.

To be able to calculate a credible range of economic abatement cost estimates for each option, we made three cost scenarios: 2.5%, mean, and 97.5%, which represents the 95% confidence interval based on estimated uncertainty in technical assumptions, energy prices and discount rate. Table 5.3 shows the technical assumptions for the option 'wind on-shore'. The complete set of assumptions for all electricity options can be found in Annex A.4.

Table 5.3   *Assumptions underlying cost estimates of on-shore wind power excluding externalities*

option	fuel		investment		fuel efficiency		O&M var	O&M fix	Load factor	lifetime	CEF	References
			2010	2020	2010	2020						
			€/kW	€/kW	%	%	€/MWh	€/kW	%	yr	tCO2/MWh	
wind on-shore	low		726	657			3.0	28.6	23%	10	0	NEA/IEA, 2005
	mean		887	700			5.4	35.7	29%	15	0	CPB/ECN, 2005
	high		1026	682			7.2	42.8	34%	20	0	Menkveld, 2004

Note: CO_2 emission from the wind turbine construction phase are not taken into account in the emission factor.

Table 5.1 gives a brief overview of the variables on which our cost estimates are based. Most assumptions are taken from NEA/IEA (2005), as this is the most broad and up-to-date study on electricity in Europe. These data are compared to, checked against and complemented by results from other studies, including CPB/ECN (2005), ECN (2005) for wind and biomass, Lako (2004) for coal and IGCC, and Menkveld (2004).

Using standardised cost calculations based on net present value using a range of discount rates, we obtained generation cost of electricity. Table 5.4 shows the result of these calculations using social discount rates (i.e. varied between 3 and 5%). The range between the 2.5% and the 97.5% values show the 95% confidence interval of the production cost, arising from the uncertainty in all assumptions.

Table 5.4   Calculated electricity generation cost (using social discount rates).
CEF: CO₂ emission factor

Option	CEF	Production cost		
		2.5%	Mean	97.5%
	tCO₂/MWh	€/MWh	€/MWh	€/MWh
PCC	0.85	24	28	30
CCGT (gas-based)	0.37	34	40	46
Wind on-shore	0	38	50	62
IGCC (coal-based)	0.67	28	32	35
Biomass co-firing PCC	*0.1*	58	66	77
Nuclear LWR/EPR	0.05	21	24	26
CHP (gas-based)	0.30	34	39	60
PCC + CCS	0.085	47	66	83

For biofuels, the following table shows assumptions.

Table 5.5   Cellulose-based biofuel assumptions

Future biofuel		2.5%	Mean	High
Oil price	$2003/bl	40	49	65
	€/GJ	5.7	7.2	9.4
Biofuel cost	€/GJ	7	14	21
Biofuel CO₂ saving	%	70%	80%	90%

5.3.6   *Energy supply security externalities*

In an attempt to account for energy supply security externalities, we have used an energy supply risk premium, as explained in Section 4.7. The justification for using a premium for energy supply security risks with regard to oil and natural gas is given in section Section 4.7. In this report, we use user-defined low, mean and high values in €/GJ for 2010 for both oil and natural gas. E.g. 1 €/GJ for oil would correspond to external cost of approximately 7 $per barrel. This reflects the inherent risk associated with importing energy from large distances at volatile prices and unsure availability, as well as the depletion of energy sources. It should be noted that this risk factor is highly uncertain even in the mid-term. E.g. if oil production would peak earlier than is generally estimated by the IEA and oil-producing nations, the value would increase strongly.

Table 5.6 shows the base year (2003) values for the risk premium for oil and gas, in three estimates: conservative, central and maximum[51]. These values increase over time with the same factor as the fuel prices to which they relate, e.g. the conservative oil supply premium increases with 1% per year, the central premium with 2% per year (see Table 5.2).

In our approach however these values are not simply translated into monetary savings. As explained in Section 4.7, the interaction with the oil and gas prices is important. At the highest oil and gas prices scenarios (the 97.5% column in Table 5.2), we have assumed that the supply risk is already adequately reflected in the price of fuels. This highest price scenario is also the maximum price trajectory for the remaining two scenarios: the mean oil or gas price plus the 'central' risk premium (both in €/GJ) in a certain year cannot exceed the highest price scenario. In Annex A.5 this is shown in two graphs for oil and gas. It can be seen that for the 'mean' price + 'central' premium scenarios, the maximum price is reached; for oil by 2017 and for gas by 2024.

Table 5.6   Assumptions regarding security of supply externalities in the numerical example

Risk premium 2003	Conservative	Central	Maximum
€/GJ oil	0.5	1	3
$/barrel of oil saved	3	7	21
€/GJ gas	0.4	1	2
€ct/m3	1.3	3	6.3

Note: See Section 4.7 for explanation. In the 'maximum' column, it is assumed that the supply premium is adequately reflected in the high (97.5% scenario) oil and gas prices, and therefore is equal to zero if the externality is calculated separately.

An alternative to this approach would have been to adopt a construction where the price of oil or natural gas increases by a constant or increasing percentage. In fact, an energy supply security risk premium has a simi-

---

[51] 'Maximum' should not be read as an absolute maximum value, but as the highest estimate of the energy supply premium assumption in our calculations.

lar effect on end use prices as a ("energy", "climate change") tax on oil and gas.

5.3.7 Air pollution externalities

In order to give a socio-economic valuation of the costs associated with GHG abatement measures, all (avoided) externalities should be incorporated. As discussed in chapter 4, including all costs and benefits in monetary terms is extremely challenging and will inevitably entail subjective parameter choices (such as valuation of health damage).

Compared to other externalities, impacts of air pollution arising from energy production and consumption are probably the most studied, notably in the context of the ExternE project. To properly value air pollution externalities, we need data on:

- Emission factors in the baseline and GHG option case

- Marginal abatement costs of the air pollutants

- Damage factor

As explained in Section 4.5, the latter two variables represent two different approaches to external cost calculation. The marginal abatement approach typically yields (much) lower cost estimates compared to the damage approach. We have prepared cost estimates based on both approaches to give insight into the range of costs. Uncertainties on emission factors are relatively low compared to damage factors (see Section 4.5), and to a lesser extent the marginal abatement cost for air pollutants. Uncertainty on damage factors being orders of magnitude higher than marginal abatement costs, we rely mostly on the marginal abatement cost approach.

Most data on emission factors of electricity production and damage factors are based on Rabl and Spadaro (2000), as used in the ExternE methodology. Abatement cost for air pollutants are taken from the RAINS model for a set of European countries (quoted in Rabl et al, 2005). These represent stationary sources (although country differences can be substantial, for the present project these values are deemed sufficiently accurate). We take these air pollutant abatement cost values also as estimates for the transportation sector, but this is an underestimate as costs are likely to be much higher. Other data sources are Lako (2004), IIASA (2005), and European Environmental Bureau (2005). Table 5.7 shows all assumptions used for the air pollution externality calculations.

Table 5.7 Air pollution externality assumptions (the bottom four rows show the final valuation assumptions in the calculations)

Emission factors	tCO_2/MWh	kg PM/MWh	kg NO_x/MWh	kg SO_2/MWh
PCC state of the art	0.85	0.2	2	1
Gas state of the art	0.35	0	0.2	0
IGCC	0.62	0.032	0.25	0.13
PCC CCS	0.085	0.037	1.2	0.68
Biomass co-firing	0-0.2	0.032	0.6	
	tCO_2/GJ	gPM/GJ	gNOx/GJ	
Gasoline car with catalyst	0.072	8	555	
Future biofuel	0.014	8	555	
Hydrogen (gas + CCS)	0.01	-	-	

Air pollution damage factors	
€/kg PM	15.4
€/kg SO_2	10.2
€/kgNO_x	16

Years of life lost due to PM	
€/YOLL	83000
€/YOLL (acute mortality)	155000

Air pollution externalities	Conservative	Central	Maximum
€/kgNO_x	0.5	7	16
€/kgSO_2	0.3	5	10
€/kgPM	0.3	1	15

For each tonne of CO_2 that is saved when an abatement option is implemented instead of its reference technology, the emissions of air pollutants are increased or decreased. The change in emissions is multiplied by the monetary factors (€/kg) to obtain the externality per tonne of CO_2 saved (see Section 5.7.1 for an example calculation).

5.3.8 Employment externalities

Net effects on employment are difficult to quantify, while often double counting is resorted to by only including direct employment gains. Plausible estimation of even the sign of the impact (positive or negative) can sometimes be debated. Therefore we will not report quantified impacts of the GHG mitigation option. Only an indicative qualitative change in employment due to the measure will be given. This should be read as compared to a similar investment in GHG mitigation in large-scale fossil fuel production or nuclear power generation.

5.4   Cost estimates based on conventional economic analysis

For most of the selected CO_2 reduction options, we have carried out cost calculations. The CO_2 abatement cost of a certain option refers to the additional cost of this option (compared to the reference) divided by the amount of CO_2 saved (compared to the reference option). To determine the cost of the options, net present value calculations are carried out. The following example shows how the mean abatement cost for wind on-shore replacing PCC was calculated:

$$Abatement \ \ Cost = \frac{ElecCost_{wind} - ElecCost_{PCC}}{CEF_{PCC} - CEF_{wind}} = \frac{50-28 \ \ [€/MWh]}{0.85-0 \ \ [tCO_2/MWh]} = 30 \ \ [€/tCO_2]$$

where CEF is the CO_2 emission factor in tCO_2/MWh.

Table 5.8 shows the GHG abatement cost estimates for the options. The figures in the column 'mean' show the value of the abatement cost of the option when it is implemented instead of the reference option, e.g. wind power on-shore instead of a coal-fired power plant, average across the EU-25 in 2010. The mean figure therefore does not apply to a specific baseline and does not reflect differences in site-specific conditions. In practice for each option, a cost curve would apply, that would show how the costs increase as a function of the cumulative capacity (of e.g. wind power) implemented. In other words, the total potential (not assessed in this study) will not be harnessed at a certain cost level, but will increase with the share of the potential implemented. The columns '2.5%' and '97.5%' reflect the uncertainty in our calculations as explained in Section 5.2.3.

Most 'mean' abatement cost values are below 50 €/tCO₂. The results clearly show the impact of different assumptions, which is the result of uncertainty in underlying parameter values such as discount rate, investment cost or current efficiency, and future parameters such as energy prices and decrease of investment cost.

Table 5.8   Overview of cost estimates (discount rate 5-10%)

Option	Reference	Economic cost		
		2.5% €/tCO$_2$	Mean €/tCO$_2$	97.5% €/tCO$_2$
Nuclear	CCGT	-70	-30	4
CHP	CCGT	-117	-30	35
Insulation	Oil/no insulation	-83	-22	106
Insulation	Gas/no insulation	-83	-22	106
Nuclear	PCC	-11	-1	9
CHP	PCC	-3	12	27
Heating efficiency	St. gas boiler	-200	23	50
Wind on-shore	PCC	11	30	49
IGCC (coal-based)	PCC	-5	31	66
CCS industry	No CCS	10	35	60
Wind on-shore	CCGT	-10	46	95
PCC + CCS	PCC	23	50	76
Biomass co-firing PCC	PCC	40	52	65
PCC + CCS	CCGT	16	105	184
Biomass co-firing PCC	CCGT	64	115	159
Biofuel (2nd gen.)	Gasoline/diesel	21	118	219

Note: The 2.5%-mean-97.5% figures are calculated based on assumptions varied according to their uncertainty distributions, with the discount rate between 5 and 10% (mean 8%), see also Section 5.2.3.

For the residential and service sector, the low and mean cost esti-mates are taken from Ecofys (2005), with the low cost figure corresponding to the implementation of insulation measures coupled with renovation based on the end-user approach, and averaged across three climatic zones. The mean estimate refers to the same measures and approach, but imple-mented independent of renovation. The high estimate is calculated using cost and energy savings from ECN/MNP (2005) for The Netherlands in 2020, based on the national (social) cost approach. As building efficiency in The Netherlands is relatively high in comparison to many other EU mem-ber states, these figures are taken as the high estimate.

The relatively large range of costs estimates for the buildings sector originates from the fact that different approaches are used in the low, mean and high values. Ideally harmonised assumptions and cost approaches would be used in the cost estimates. As these were not available in this study, however, we assume that the reported values give a credible (but large) range of costs.

5.5   Impact of applying social discount rate

Table 5.9 shows how economic cost figures change with the applied discount rates, for the power sector options. The other options are not shown as no NPV calculations were carried out for these. Columns 3 to 5 show the results of calculations using relatively high discount rates (as applied in most standard calculations). These columns repeat, for the reader's convenience, the results given in Table 5.8.

Column 6 to 8 give results of abatement cost calculations using lower discount rates, as justified in a socio-economic analysis. The 2.5%, mean and 97.5% cost values again reflect uncertainty in assumptions.

The calculations in columns 3 to 5 respectively 6 to 8 are based on assumptions with the same uncertainty, the only difference being the range of discount rates. Therefore, the range of the results will be shifted and can be compared to gain insight into the effect of applying a lower discount rate range.

The change in the mean abatement costs arising from the change in discount rates is shown in the rightmost column. It can be observed that this change has a significant impacts on the abatement cost. For instance, when CCGT is the reference option, capital-intensive options such as nuclear and wind on-shore, but also biomass co-firing and CHP gain significantly if a lower discount rate is used. As PCC generation cost decreases more compared to CCGT, some options do relatively less well when compared to PCC as a reference option, such as CHP (gas-based).

Table 5.9   Impact of change in economic abatement cost of CO$_2$ options

1	2	3	4	5	7	8	9	12
option	reference	economic cost (discount rate 5 - 10%)			economic cost (discount rate 3 - 5%)			change (mean)
		2.5%	mean	97.5%	2.5%	mean	97.5%	
		€/tCO$_2$	€/tCO$_2$	€/tCO$_2$	€/tCO$_2$	€/tCO$_2$	€/tCO$_2$	€/tCO$_2$
nuclear	CCGT	-70	-30	4	-87	-51	-22	-20
CHP	CCGT	-117	-30	35	-118	-5	66	25
nuclear	PCC	-11	-1	9	-12	-5	3	-4
CHP	PCC	-3	12	27	4	21	38	10
wind on-shore	PCC	11	30	49	8	26	44	-4
IGCC (coal-based)	PCC	-5	31	66	-4	25	53	-6
wind on-shore	CCGT	-10	46	95	-30	27	77	-19
biomass co-firing PCC	PCC	40	52	65	40	97	145	45
PCC + CCS	CCGT	16	105	184	1	92	172	-13
biomass co-firing PCC	CCGT	64	115	159	40	97	145	-17
Biofuel (2nd gen.)	gasoline/diesel	21	118	219	20	118	218	0

Note: The 2.5%-mean-97.5% abatement cost figures are calculated based on assumptions varied according to their respective distributions (with 2.5%, mean and 97.5% values listed in Annex A.4), with the discount rate between 5 and 10% in column 3,4 and 5, and between 3 and 5 in column 6-8).

Figure 5.2 depicts the uncertainty in abatement cost results, based on our low, mean and high assumptions and the uncertainty analysis. We can observe that the uncertainty in outcomes is substantial for several options, up to 200 €/tCO$_2$. This can be explained by uncertainty in key assumptions, and sensitivity of the outcomes to these variables, such as fuel prices (for both the options and the reference), lifetime of the technology, specific investment cost, etc. This is especially valid for biofuel, biomass co-firing, wind on-shore and CHP. For the options in the residential sector the uncertainty reflects the range in abatement cost as found in the literature sources. This also shows that the outcomes are very dependent on the assumptions and cost approach chosen.

Figure 5.2   Uncertainty intervals (95% confidence intervals) resulting from economic CBA analysis of climate change mitigation options

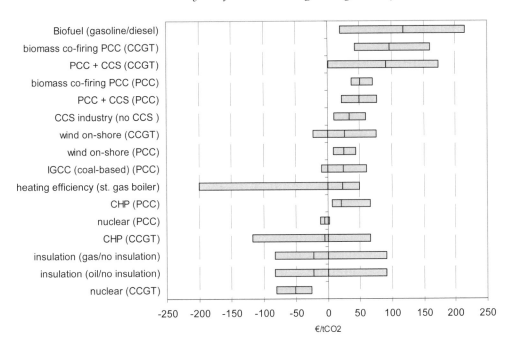

It should be noted that the bandwidths here presented should not be regarded as being an accurate representation of the actual uncertainty in abatement costs, but merely as the result of uncertainty in our assumptions, which are based on a relatively small set of literature and expert judgement. In a more elaborate assessment, bandwidths for some options are likely to be smaller due to better knowledge about technology assumptions, i.e. better data. To some extent however, uncertainty is inherent to abatement cost studies. This is due to spatial variability, inherent uncertainty about e.g. investment cost (wind, nuclear, biofuel) and baseline uncertainty.

5.6   Impact of including externalities in cost calculations

Table 5.12 shows abatement costs for the climate change mitigation options, using a social discount rate. Column 3 to 5 summarise our results for the economic cost calculation, i.e. excluding externalities, as presented in Section 5.5, at the discount rate range of 5-10%. However, the options for which no calculations have been carried out are also included; literature estimates are taken.

The options are ranked according to the mean values for abatement cost (column 6). Columns 7 to 9 indicate abatement costs when the externalities for air pollution and energy supply security are taken into account. In Section 5.7 we discuss the calculation of the externality value for each option. In the last two columns, the change in (mean) abatement costs and the change in the ranking thereof, as a result of including externalities, are shown.

In general, including externalities decreases the abatement cost of the climate change mitigation option. Only in two cases, i.e. biomass co-firing and PCC+CCS compared to CCGT, the option yields negative benefits (increases the abatement cost). Therefore, the 'low' value of the cost including externalities is calculated using the highest externality values, as in this case the option yields the most 'benefits' compared to the reference, and thus the lowest abatement cost. The 'high' cost value corresponds to the most conservative externality assumptions, yielding the lowest benefits and therefore highest cost. For consistency, this approach has also been taken for the two options biomass co-firing and PCC+CCS, as the 'low' abatement costs for the options are calculated using high externality values and the 97.5% CCGT electricity cost, which is calculated using the highest gas price scenario (see Section 5.3.6 for clarification on the approach for energy supply security externalities).

Table 5.10 Ranking of options, based on economic cost excluding and including AQ and ESS externalities

1	2	3	4	5	6	7	8	9	10	11	12
		economic cost				social cost				change	
option	reference	2.5% $€/tCO_2$	mean $€/tCO_2$	97.5% $€/tCO_2$	rank (mean)	2.5% $€/tCO_2$	mean $€/tCO_2$	97.5% $€/tCO_2$	rank (mean)	rank	$€/tCO_2$
nuclear	CCGT	-87	-51	-22	1	-94	-60	-28	3	▶	-9
insulation	oil/no insulation	-83	-22	91	2	-164	-62	80	2	=	-40
insulation	gas/no insulation	-83	-22	91	3	-93	-33	82	4	▶	-11
CHP	CCGT	-118	-5	66	4	-126	-13	59	6	▶	-8
nuclear	PCC	-12	-5	3	5	-78	-29	1	5	=	-24
CHP	PCC	4	21	38	6	-83	-9	65	7	▶	-31
heating efficiency	st. gas boiler	-200	23	50	7	-210	12	41	9	▶	-11
IGCC (coal-based)	PCC	-4	25	53	8	-268	-68	55	1	◀	-93
wind on-shore	PCC	8	26	44	9	-53	3	43	8	◀	-23
wind on-shore	CCGT	-30	27	77	10	-34	19	74	10	=	-8
CCS industry	no CCS	10	35	60	11	10	35	60	12	▶	0
PCC + CCS	PCC	24	50	75	12	22	36	38	13	▶	-14
biomass co-firing PCC	PCC	38	51	64	13	-13	33	53	11	◀	-18
PCC + CCS	CCGT	1	92	172	14	71	112	172	16	▶	20
biomass co-firing PCC	CCGT	40	97	145	15	86	107	159	15	=	10
Biofuel (2nd gen.)	gasoline/diesel	20	118	218	16	21	99	204	14	◀	-20

Note: Input distribution for discount rate between 3 and 5% (triangular distribution with mean 4%) for both sets of calculations in columns 3 to 5 and 7 to 9.

If we look at the change in the ranking, it appears that several options improve their cost-effectiveness considerably: IGCC, insulation (reference oil), wind and biomass co-firing (although this is not necessarily reflected in the ranking). Another important observation is that for almost all options abatement costs decrease significantly, by up to 93 €/tCO$_2$ taking the mean externality values.

Some notes need to be made to this way of presenting the results

- the uncertainty in the economic cost calculations is much larger than the externalities;

- the uncertainty in the externality calculations is not visible anymore, which is an important factor to be taken into account;

- it can be debated whether the changes in ranking are significant if compared to the uncertainty ranges;

- only two externalities are taken into account, which means that any other costs or benefits potentially significant are neglected here.

However, this section shows that the impact of including externalities in abatement cost can be significant. In Section 5.7 the externalities will be discussed in more detail.

5.7 Cost including (avoided) externalities

In order to give a picture of the cost faced by society as a whole, all externalities or avoided externalities should be included in cost calculations. This section aims to quantify these externalities - if possible in monetary terms - and to present a best-effort cost estimate.

5.7.1 Energy and industry

Several major GHG emission reduction options in power production - wind, biomass, IGCC and nuclear - all achieve other policy goals such as energy security and air pollution reduction, though to different extents. We have calculated the externalities for each unit (MWh) of electricity produced from these technologies as well as from the reference technologies PCC and CCGT, which is shown in Table 5.11.

Table 5.11 Externalities of air pollution and energy supply security per unit of electricity, technology-wise

option	CEF tCO2/MWh	Cost (excluding externalities) 2.5% €/MWh	mean €/MWh	97.5% €/MWh	air pollution low €/MWh	mean €/MWh	high €/MWh	Risk premium low €/MWh	mean €/MWh	high €/MWh	Externalities total low €/MWh	mean €/MWh	high €/MWh
PCC	0.85	24	28	30	1	19	53	0.7	1.4	0	1	19	53
CCGT (gas-based)	0.37	34	40	46	0.1	1.5	4				1	3	4
wind on-shore	0	38	50	62							0	0	0
IGCC (coal-based)	0.67	28	32	35	0.1	2	7				0	2	7
biomass co-firing PCC	0.1	58	66	77	0.3	6	16				0	6	16
nuclear LWR/EPR	0.05	21	24	26							0	0	0
CHP (gas-based)	0.30	34	39	60	0.1	1.3	3	0.6	1.1	0	1	2	3
PCC + CCS	0.085	47	66	83	0.5	9	24				1	9	24

The air pollution externality is determined by a NPV calculation, using the 97.5% values of the lifetime for the plant and discount rate (5%) for the low externality value. For the the high externality value, the 2.5% values for the lifetime and discount rate are used; mean lifetime and discount rate for the mean externality. For each year the electricity production (MWh) is multiplied by the emission factors (kg/MWh) and the cost factors (€/kg) for PM, NO$_x$ and SO$_2$.

$$AP_{PCChigh} = \frac{NPV\left(3\%, \displaystyle\sum_{LTlow}(Elec\Pr od * (EF_{PM} * Cost_{PMhigh} + EF_{NOx} + Cost_{NOxhigh} + EF_{SOx} * Cost_{SO2high}))\right)}{NPV\left(3\%, \displaystyle\sum_{LTlow} Elec\Pr od\right)} \quad [\text{€} / MWh]$$

where:

$AP_{PCChigh}$	air pollution externality for electricity from PCC
NPV	Net Present Value
ElecProd	Annual electricity production (MWh)
3%	Discount rate used
EF	emission factor (kg/MWh)
$Cost_{PMhigh}$	valuation (high) of external cost of particulate matter emissions (€/kg) (similar for SO_2 and NO_x)

For the energy security externality, the approach is analogous: instead of an emission factor per MWh, the required input of natural gas (GJ/MWh) is multiplied by the cost factor (€/GJ) per year as set out in Section 5.3.5, after which the NPV of the total cost across the lifetime is calculated per MWh.

In Table 5.11, the values for 'externalities total' are calculated simply by adding the low value for AP to the low value for ESS. Using the total externality values the social abatement cost (i.e. including externalities) in Table 5.10 were calculated. For example, to calculate the 'low' social abatement cost when wind on-shore replace CCGT electricity:

$$SAC_{CCGT \to wind(low)} = \frac{ElecCost_{wind2.5\%} + Ext_{windhigh} - ElecCost_{CCGT97.5\%} - Ext_{CCGThigh}}{CEF_{CCGT} - CEF_{wind}} [€/tCO_2]$$

where:

$SAC_{CCGT\text{->}wind(low)}$	social abatement cost for wind replacing CCGT, low estimate
$ElecCost_{wind2.5\%}$	2.5% estimate for wind electricity production cost [€/MWh]
$ElecCost_{CCGT97.5\%}$	97.5% estimate for CCGT electricity production
$Ext_{windhigh}$	Externality estimate in €/MWh, high estimate
CEF	CO_2 emission factor [$kgCO_2$/MWh]

The basis for quantifying employment benefits is too weak and case-specific to be included here. Quantification of possible employment effects requires huge input-output modelling efforts far beyond the scope of the present study. Broadly speaking, capital-intensive, low-expense technologies are not likely to result in major employment benefits, contingent on whether most of the plant installations are imported or produced within the EU. As for fuel inputs, it also matters whether these are produced inside or outside the EU. As for biomass fuels a great caveat is in order. Many authors claim huge employment benefits from energy cropping (e.g. Faaij, 2006, but also several publications of the European Commission). Yet a solid analysis is required to find out what type of activities energy cropping replaces (presumably agricultural activities), whether these are more or less labour-intensive than the replaced activity, and whether they are more or less subsidized by the EU consumers or the public sector. In the fortunate event that labour-intensive, non-subsidized energy cropping activities replace labour-extensive agricultural activities heavily subsidised by the Common Agricultural Policy, sizable employment benefits can be reaped indeed. However, given increasing tax exemptions, such as excise rate reductions or outright exemptions granted to the production of biofuels and other bio-energy applications, these benefits should not be taken for granted.

For biomass, explicit attention should be given to the sustainability aspect of its production. E.g. if biodiversity is negatively affected in tropical regions by palm oil plantations, this impact should be taken into account as a cost or negative benefit.

CO_2 capture and storage applied at a coal-fired power plant achieves little ancillary benefit, except reduction in PM emissions and possible synergies with technological development of (low-carbon) hydrogen production. When CO_2 capture and storage is applied in (coal-using) industries or power plants, however, the main direct co-benefit is a large reduction in particulate matter emissions (up to 80%). Indirectly, application of CCS enables industries to continue using coal as a fuel within a stricter climate change policy environment. As coal is expected to be an important source of relatively cheap energy in the coming decades (IEA, 2005), this is a major issue. Coal users and producers are aware of this and are looking for ways that enable them to continue coal production and use in an environment-friendly fashion (IEA/OECD, 2005).

Another indirect benefit consists of the importance of CCS deployment in the longer-term energy policy. CCS is likely to be a crucial technology in the transition towards a hydrogen-based energy system. This of course is based on the premise that (international) climate policy will remain and imposes stricter CO_2 emission limits on (European) countries (Bruggink, 2005). Therefore, in the light of long-term policy, stimulating CCS can be seen as a prudent strategy.

Deployment of combined heat and power has - depending on the primary energy source used - important benefits for air quality and security of energy supply. If it is applied at gas-fired generation facilities, the security of energy supply improves and NO_x emissions are reduced. In case of coal-based capacity, reduction in air pollutant emissions are the most important benefits (see Table 5.11).

As stated above, the indirect benefit of being able to operate based on fossil fuels in a climate-constrained policy environment should be acknowledged. Even CHP is likely to an important technology in achieving this.

5.7.2 Transport sector

Biofuels substituting conventional fuel such as gasoline is widely seen as an important option to reduce dependency on (imported) oil. It therefore contributes significantly to enhancing security of energy supply, also for the reason of diversification. Every barrel of oil saved is by definition at the margin, which typically comes from the most risky region, implying large benefits.

Employment in the agriculture may increase significantly due to large-scale biofuel utilisation. Wakker et al. (2005) estimate that nearly 30,000 person-years per annum in Poland, and more than 10,000 in Hungary, France and Spain, will be generated in the agricultural sector when the potential is used. Again, it is stressed that a total employment analysis is in order to justify genuinely robust statements in this respect.

On the other hand, the source of biomass is important to determine local benefits and negative impacts. Large palm oil plantations in tropical nations often threaten high-biodiversity forests. IEA (2003) notes that biofuel production may produce net environmental benefits under the right conditions, however. Biofuel utilisation is not likely to have a significant effect on urban air quality (IIASA, 2005), however this needs to be further explored (see also IEA (2003)).

Fuel switch from gasoline or diesel to natural gas or liquefied petroleum gas is one of the most important options to improve air quality. Several megacities in developing countries, e.g. Delhi, have introduced such a fuel switch policy. However, as smog is also an important environmental problem in urban centres in industrialised countries, this option may improve air quality in Europe as well (Kok & De Coninck, 2004).

Hydrogen as a fuel, produced from coal (or biomass) with CO_2 capture and storage would result in major benefits[52] for:

- (urban) air quality and reduced acidification

- energy security

---

[52] These benefits probably cannot be called 'ancillary' as they are likely the primary reasons to promote this technology, particularly in the US and developing countries, but also in Europe.

Table 5.12 Summary table for externality outcome for the transport sector

Option	Reference	Economic cost €/tCO$_2$			Air pollutions benefits						ESS benefits					
					Average emission (kg/tCO$_2$)			€/tCO$_2$			€/tCO$_2$			€/tCO$_2$		
		2.5%	Mean	97.5%	kgPM	kgNOx	kgSO2	Low	Mean	High	Low	Mean	High	Low	Mean	High
Biofuel (2nd gen)	Gasoline	21	118	215				0	0	0	11	20	0			
Hydrogen fuel cells	Gasoline				0.12	9		4	44	139	14	22	0			

Proposed Euro V standards for particulate and NO_x emissions in 2008 are 2.5 and 80 g/km respectively. Applying hydrogen fuel cells will result in zero emissions in the urban environment, abating 10 and 400 tonnes of PM and NO_x emissions if applied at a scale that 1 $MtCO_2$ would be avoided by CO_2 free hydrogen. This can be translated into avoided abatement cost of € 0.2 - 1 million, or 7 million in avoided loss of life.[53]

For energy security, the biofuel and hydrogen measures have comparable benefits in the order of 10 - 20 €/tCO_2 using the ESS assumptions in section 5.3.6.

5.7.3 Residential and services sector

Major benefits from the discussed options in the residential and services sector, insulation and heating efficiency, include energy security and enhanced comfort of living. This is based on the assumption that mostly natural gas is used for heating.[54]

Enhanced comfort of living by insulation measures has several aspects:

- reduced noise
- decrease in condense or humidity
- reduced cold air flows

At a € 0.4 - 2/GJ risk premium for natural gas, energy security benefits of 23 PJ gas reduction, required for 1 $MtCO_2$/yr reduction, would amount to € 9 - 46 million. If we assume that energy savings come from a reduction in the use of petroleum products, energy security benefits add up to € 9 - 54 million.

Implementation of the Energy Performance for Buildings Directive would have 'moderate' net employment benefits in the order of 10,000 to 100,000 jobs in Europe, according to Ecofys (2005). Investments required for the proposed energy efficiency measures are € 10-25 billion annually, which

---

[53] This figure is based on a power plant damage factor and hence, if corrected for urban emissions the figure will be higher.

[54] In some Nordic countries, heating systems may use electricity. In this case benefits depend on which fossil fuel source is used for marginal power production.

corresponds to approximately 1-3% of the total construction investment in the EU.

In addition, the energy efficiency measures discussed in this report have a significant potential to reduce air pollution. Especially for NO_x, where the household and service sectors contribute the lion's share, this is a very significant benefit. Assuming 130 $MtCO_2$ potential savings (2010, Joosen & Blok (2001), see Section 3.4), this can be translated into reductions in energy consumption and emissions, shown in Table 5.13. The figures for gas and petroleum should be read separately, e.g. they reflect the air pollution reduction if all the measures that are implemented reduce gas consumption. A combination of savings of gas and petroleum (in total about 2000 PJ) is the most likely real outcome; the shares are however unknown, and therefore they are shown separately. Table 5.14 summarises the externality findings for the buildings sector.

Table 5.13 Air pollution reduction at 130 $MtCO_2$/yr reduction measures.

Energy source	PJ saved (maximum)	PM		NO_x		SO_2	
		EF	ER	EF	ER	EF	ER
		g/GJ	kt	g/GJ	kt	g/GJ	kt
Natural gas	1800	0	0	20-50[55]	46-115	0	0
Petroleum	2300	20-50	36-90	61[56]	179	218	501

EF: emisson factor; ER: emission reduction (in kilotonnes).

---

[55] Kroon et al., 2005.

[56] DoE, 2000.

Table 5.14 Externalities for residential sector options

Option	Reference	Economic cost €/tCO₂			Air pollutions benefits						ESS benefits		
					Average emission (kg/tCO₂)			€/tCO₂			€/tCO₂		
		2.5%	Mean	97.5%	kgPM	kgNOx	kgSO2	Low	Mean	High	Low	Mean	High
Insulation	Gas/no ins.	-83	-22	91		0.6		0.3	4	10	9	17	0
Insulation	Oil/no ins.	-83	-22	91	0.7	2.5	3.0	2	33	81	9	16	0
Heating efficiency	St. gas boiler	-200	23	50		0.6		0.3	4	10	9	17	0

5.7.4   Technological learning

The calculations in the previous sections were carried out assuming the technologies are implemented and operational in 2010.[57] If 2020 is taken as the starting year instead, the results may change (as explained in 4.3). Table 5.15 shows how (mean) abatement costs for wind on-shore may change as a result of the decreased specific investment costs when the technology is implemented in the year 2020.

Table 5.15 Example calculation showing impact of technological learning
(abatement cost for wind on-shore replacing PCC in 2020).

	Investment ($€/kW_e$)		Elec cost ($€/MWh$)		Abatement cost ($€/tCO_2$)	
	2010	2020	2010	2020	2010	2020
Wind on-shore	887	700	50	37	27	12
PCC	1100	1067	28	27		

Table 5.16 shows results for all technologies, based on calculations similar to Table 5.15. We note that these figures are only given as examples, as these results arise from a simple technological learning model based on straightforward assumptions. Also, these are costs excluding externalities. Including these would not change the relative results.

This example shows that some technologies, notably wind power and biomass co-firing, may be more cost-effective in the future. However, this is based on the assumptions of technological learning, which in turn assumes a certain rate of implementation of the technology. This therefore depends on the extent to which the technology is stimulated in the short and me-dium term.

---

[57] Except for 2nd generation biofuel, which are assumed to be implemented from 2020.

Table 5.16 Changes in abatement cost using 2020 as starting year

1	2	3	4	5	6	7	8	9
option	reference	economic cost (2010)			economic cost (2020)			change
		2.5%	mean	97.5%	2.5%	mean	97.5%	(mean)
		€/tCO$_2$	€/tCO$_2$	€/tCO$_2$	€/tCO$_2$	€/tCO$_2$	€/tCO$_2$	€/tCO$_2$
nuclear	CCGT	-87	-51	-22	-133	-71	-30	-20
CHP	CCGT	-118	-5	66	-160	-9	78	-4
nuclear	PCC	-12	-5	3	-12	-4	3	1
CHP	PCC	4	21	38	10	33	61	12
IGCC (coal-based)	PCC	-4	25	53	-12	19	48	-7
wind on-shore	PCC	8	26	44	-3	12	27	-14
wind on-shore	CCGT	-30	27	77	-96	-25	31	-52
biomass co-firing PCC	PCC	38	51	64	38	51	64	0
biomass co-firing PCC	CCGT	40	97	145	-16	72	135	-26

Note: input distribution for discount rate: 3-5%

5.7.5 Should cost of inaction be included?

In a fully comprehensive cost-benefit analysis of climate change mitigation options, the avoided damage costs due to climate change should be included. We however chose not to include them, for 1) it is not relevant for a *comparison* between CO_2 options and 2) the benefits accrue to the global environment while the cost is borne by the EU. Moreover, the point was made that assigning a figure to the damage costs is surrounded by large uncertainties and subjective choices (see Section 4.8). Cost-effectiveness analysis of climate change mitigation options avoids becoming embroiled in the debate on the valuation of climate damage costs. Even so, this report shows that a social perspective of non-climate ancillary costs and benefits can be resorted to in conducting such analysis.

6. Concluding observations

This exploratory study has introduced a new standard framework for undertaking cost-benefit analyses of climate change mitigation options for public policy purposes. Ubiquitous adoption of such a framework will greatly improve comparability of cost information between distinct options and member states. Moreover, option cost information derived by application of the proposed framework will make more appropriate allowance for (often significant but longer term) co-benefits outside the realm of climate change.

The essentials of the proposed standard framework for social cost-benefit analysis of distinct climate change mitigation options for public policy purposes are captured by the following broad guidelines:

1. *Check the interactions of the options reviewed and make sure that options retained for policy implementation purposes are not incompatible with each other.*

2. *Use efficiency prices (i.e., by and large, market prices net of taxes and subsidies) as the point of departure for cost-benefit analysis from a societal point of view.*

3. *Analyse explicitly the context-specific suitability of applied discount rates without "automatically" applying discount rates used by authoritative economic development analysis and planning bodies.*

4. *Show quantitatively uncertainties surrounding resulting key figures regarding mitigation cost per option.*

5. *Make serious efforts to quantitatively include major external costs and benefits in resulting key figures.*

Starting out from a conventional framework, the proposed framework permits quite well to gauge successively the impact of alternative choice of discount rate and distinct externalities on resulting cost per CO_2-eq estimates.

In a numerical example the proposed framework has been demonstrated for selected climate change mitigation options. Two major co-benefits were explicitly addressed in a quantitative way, i.e. air quality (AQ) co-benefits and energy supply security (ESS) co-benefits.

To assess the social benefits of avoided SO_2, NO_x, and PM pollution cost two approaches were adopted. Firstly, the avoided abatement cost for achieving air pollution policy goals for PM, NO_x and SO_2, were projected. Second, avoided health damage costs were projected using ExternE based valuation of human life. The first approach yields lower externalities and was applied to the low and central estimates, while the latter approach was pursued for obtaining the high estimates.

For quantifying the externality of long-term energy supply security (ESS) the focus was put on the long-term supply risks of two fossil fuels, i.e. oil and natural gas. The proposed procedure for deriving the social ESS cost of oil or gas use runs as follows. For the base year a fuel-specific 'risk premium' is set. This risk premium is to reflect the social cost of oil and gas use in terms of reduced energy supply security to the extent that this is not reflected in market prices. In the 'high energy price scenario' the ESS risk is supposed to be properly reflected in the market price. Therefore, the risk premium plus the energy price cannot exceed the high price. In line with the Hotelling rule for the price trajectory over time for exhaustible resources, the premium is assumed to increase over time according to the social discount rate. The base year risk premium is inherently subjective in nature and can be set in a dialogue between scientists and policymakers in order to improve the acceptance of the social cost assumptions used. Furthermore, ESS risks seem less pronounced for uranium and, even more so, coal. Nonetheless, the ESS risk valuation approach for oil and gas use can be readily extended to uranium and coal as well.

In the numerical example the impact of changing the discount rate was gauged first. By changing the discount rate from values applied in standard contemporary economic cost calculations (5-10%) to values reflecting the social perspective (3-5%) it was confirmed that the choice of discount rate has a significant bearing on cost results. Capital-intensive options such as wind and nuclear power improve their cost-effectiveness compared to the reference (fossil) options.

Table 6.1 summarises per selected option findings of application of the proposed analysis framework in the numerical example. This table should be read in conjunction with Figure 6.1.

Table 6.1 Qualitative overview of study findings

Option	Abatement cost remarks	Remarks on benefits and issues
Wind	Cost may be lower if 1) technology cost decrease faster and 2) energy (coal/gas) prices increase	General agreement on ESS benefit due to diversification; AQ benefits due to coal baseload replacement; visual intrusion is a barrier
Biomass co-firing	Uncertainty in biomass cost	AQ: reduction in SO_2 emission; possible (small) increase in PM and NO_x
IGCC	Uncertainty in future specific investment cost	NO_x benefits large and possibly important in development of CCS and H_2 production
Nuclear	Large range of investment cost and disagreement on discount rate; waste treatment often not included in cost; replacing coal	Public opposition still large; clear benefits for AQ and ESS; large investment significant barrier; Damocles risk; cost of decommissioning uncertain
CCS (in PCC)	Uncertainty in technological learning	Indirect benefit of continued coal utilisation within CO_2 constraints; PM reduction;
CHP industry	Depends on important small differences in energy prices: uncertainty large; O&M cost uncertain	Clear benefits for AQ and ESS; better competitiveness due to efficiency
CCS industry	Cost differ for subsectors refineries, fertiliser, ethene production	Increase in fossil fuel use
Biofuel	Future biomass sources and processing cost highly uncertain	Clear benefits regarding ESS; Possibly employment benefits as well
Hydrogen fuel cells	Cost very uncertain due to infrastructural requirements	Possibly crucial option for urban AQ and ESS
Insulation	Cost-effectiveness strongly depends on cost approach (end-user or national); barriers must be taken into account; introduction also limited by housing turn-over and renovation rate	Clear benefits for AQ (particularly NO_x) and ESS (replacing gas and petroleum); most studies point to significant net employment benefits
Heating efficiency	High investment cost for small consumers	

ESS: security of supply; AQ: air quality.

Figure 6.1 gives an overview of results from applying the proposed framework to the numerical example. The 'mean' values of the economic cost (discount rate 5 - 10%) excluding externalities, and those at lower discount rate including externalities AQ and ESS shown per option, both in €/tCO₂. For readability reasons, low and high values are not displayed here and the reader is referred to Table 5.10 to appreciate the uncertainties regarding the cost results shown.

Figure 6.1 *Difference in economic cost using lower discount rates and including externalities of air pollution and energy supply security (only central values are shown).*

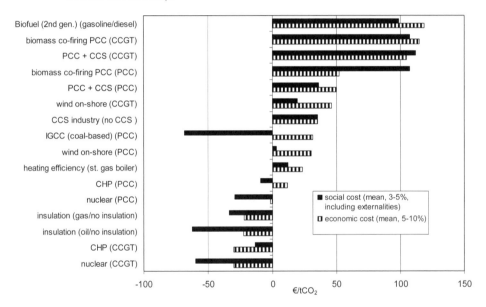

Figure 6.1 shows the remarkable differences occurring when a wider concept of 'social cost' is used when assessing climate change mitigation options. Especially IGCC, wind on-shore and nuclear power, biofuel and energy efficiency measures in the buildings sector exhibit strong benefits. Including benefits in the analysis may therefore change priorities and should be looked at carefully.

Including co-benefits of AQ and ESS made clear that the impact on cost outcome is substantial and may offset economic cost in some cases. Also the cost-effectiveness ranking of the options exhibits changes when the external costs and benefits are included. In particular IGCC (replacing PCC) and biofuel gain, but also the energy efficiency options and CHP improve their cost-effectiveness.

A more multi-facetted assessment approach than sheer social cost effectiveness is using the following criteria for prioritisation:

- cost-effectiveness (expressed in €/tCO₂-eq)
- co-benefits for other policy areas such as energy security of supply
- certainty about cost and benefits
- GHG abatement potential
- public acceptability
- ease of implementation
- no major negative and preferably positive interactions with related options.

Using both methods, the following broad picture emerges:

- Insulation is very cost-effective (potential medium) from the end-user point of view and has medium benefits for employment, energy security and air quality;
- IGCC has medium cost but high AQ benefits and contributes significantly to the (probable) long-term goal of applying CCS in such and other coal plants, and in the development of cost-effective hydrogen production;
- Biofuel has medium cost, high benefits for energy security, possibly for employment;
- Cost of CHP probably low to medium, and medium ancillary benefits;
- Nuclear power appears to be cost-effective and exhibits high benefits for AQ and ESS, but its suitability needs to be assessed in a much wider framework.

Based on the outcomes of the numerical example and additional qualitative information on the other prioritisation criteria, emanating from a literature scan, a broad classification of selected climate change options was made. Figure 6.2 below depicts this classification.

Figure 6.2 Broad classification of GHG mitigation options discussed

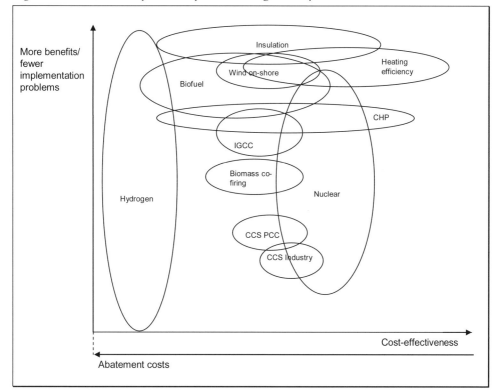

Cost uncertainties are considerable, both in the economic cost esti-mates as well as in the order of magnitude of externalities. Key factors hav-ing a high cost impact are the discount rate(s) used and energy price trajec-tories over time. *These and other cost uncertainties should be duly taken into ac-count in preparing cost-effectiveness analysis of climate change mitigation options and policy making.*

In social cost-benefit analysis, ideally an endogenous technology de-velopment approach should be used, where cost of technology is not fixed and depends on other interacting technology developments as well as pol-icy dynamics. It is important to acknowledge interaction between different options, not only on the physical impact of emission reduction estimates, but on their mutual dependence as well. For example development of CCS may depend to a certain extent on implementation of IGCC. Choices for certain technologies now may affect development of other options in the future. *It has been shown in this report that implementing technologies 10 years later may change cost-effectiveness significantly, notably for wind and IGCC. However, this is only valid provided the assumed learning rates will be really achieved. This depends to a significant extent on policy stimulation of the technology in earlier years.*

In fact, from a sustainable development and long-term energy supply security perspective, *a very high priority is warranted to put in place proper policy frameworks that foster acceleration of exhaustible-resource-saving innovations.* The climate change issue has only enhanced the urgency for human kind to accelerate sustainability-enhancing technological development.

References

Azar, C. and S.H. Schneider (2002), "Are the economic costs of stabilising the atmosphere prohibitive?", *Ecological Economics*, 42, pp. 73-80.

Azar, C. (2003), "Catastrophic events and stochastic cost-benefit analysis of climate change: An editorial comment", *Climate Change*, 56, pp. 245-255.

Bolinger, M., R. Wiser, W. Golove (20060."Accounting for fuel price risk when comparing renewable to gas-fired generation: the role of forward natural gas prices", *Energy Policy*, 34, pp. 706-720.

Boonekamp, P.G.M., J.P.M. Sijm and R.A. van den Wijngaart (2004), *Milieukosten energiemaatregleen 1990-2010. overzicht kosten en mogelijke verbeteringen in de monitoring*, ECN report ECN-C--04-040, Energy research Center of the Netherlands, Petten, NL (www.ecn.nl).

Bouwman, A.F. and D.P. van Vuuren (1999), *Global Assessment of Acidification and Eutrophication of Natural Ecosystems*, United Nations Environment Programme, Division of Environmental Information, Assessment, and Early Warning (UNEP/DEIA&EW), Nairobi, Kenya, and National Institute of Public Health and the Environment (RIVM), Bilthoven, The Netherlands.

Bruggink, J.J.C. (2005), *The next 50 years: Four European Futures,* Petten, NL.

CBS (Centraal Bureau voor de Statistiek) (2005), *Energy prices.*

CEC (2002), Directive 2002/91/EC of the European Parliament and of the Council on energy performance of buildings, Official Journal of the European Communities, L 1/65.

Chevalier, J.M. (2004), *Les grandes batailles de l'énergie,* Paris: Gallimard, Folio.

Chevalier, J.M. (2005), Security of energy supply for the European Union, Paris, 26 September.

CPB/ECN (2005), *Kosten-batenanalyse wind op zee* (in Dutch), Central Planning Bureau and Energy research Center of the Netherlands, The Hague and Petten, NL.

Daniels, B.W. (2005), *Energy-efficiency opportunities for the Dutch energy-intensive industry and refineries towards 2020,* ECN report ECN-I--05-003, Energy research Center of the Netherlands, Petten, NL.

Danish Ministry of Environment (2005), Denmark's fourth National Communication on Climate Change under the UNFCCC, Copenhagen, December.

Deyette, S. and J. Clemmer (2005), *Increasing the Texas Renewable Energy Standard: Economic and Employment Benefits* (available at www.uscusa.org).

Devarajan, S. and A.C. Fisher (1981), "Hotelling's 'Economics of Exhaustible Resources': Fifty Years Later", Vol. 19, No. 1, pp. 65-73.

DoE (2000), Emission factors for fuel combustion from natural gas, LPG, and oil-fired residen-tial water heaters (downloaded from http://www.eere.energy.gov/buildings/appliance_standards/residential/pdfs/k-2.pdf).

DoE/EIA US Department of Energy/Energy Information Administration), (2006), *Annual Energy Outlook 2006 (AEO2006) Overview*, Washington, D.C.

Donkelaar, M. ten, R. Harmsen and M.J.J. Scheepers (2004), *Advies WKK MEP-tarief*, ECN-C-04-049, European research Center of the Netherlands, Petten, NL, May.

Dougle, P.G. and R.J. Oosterheert (1999), *Case studies on energy conservation and employment in The Netherlands. Subsidy on Condensing Boilers, Subsidy on Energy Management Systems and Introduction of an Energy Performance Standard (EPN)*, ECN-C-99-060, European research Center of the Netherlands, Petten, NL.

ECN/MNP (2005), *Optiedocument energie en emissies 2010/2020* (in Dutch), Report ECN-C-05--105, European research Center of the Netherlands, Petten, NL, March.

ECN/MNP (2005b), *Reference projection energy and emissions 2005-2020*, Report ECN-C-05--089, European research Center of the Netherlands, Petten, NL, March.

Ecofys (2005), Cost-effective climate protection in the EU building stock, Ecofys report DM70086.

Ecofys (2005b), *Cost-effective climate protection in the building stock of the new EU member states. Beyond the EU Energy Performance of Building Directive*, Ecofys report DM 70067.

ECOTEC (2003), *The Impact of Renewables on Employment and Economic Growth*, ECOTEC draft report of Alterner Project 4.1030/E/97-009, Birmingham.

EIA (Energy Information Administration) (2006), *Annual Energy Outlook 2006 (AEO2006) Overview*, Washington D.C.

ExternE (2005), *Externalities of energy. Methodology 2005*, Update.

European Commission (2003), *World energy, technology and climate policy outlook 2030*, EUR 20366, Luxembourg.

European Commission (2004), *European energy and transport scenarios on key drivers*, Luxembourg.

European Commission (2005a), *Winning the Battle against Global Climate Change*, Communication from the Commission to the Council, the European Parliament, the European Economic and Social Committee and the Committee of the Regions, COM (2005) 35 final.

European Commission (2005b), *Doing more with less. Green paper on energy efficiency* (europa.eu.int).

European Environmental Bureau (2005), Environmental Policy Handbook. Chapter IV.3. Air (http://www.eeb.org/publication/chapter-4_3.pdf).

Ezzati, M., R. Bailis, D.M. Kamen, T. Holloway, L. Pirce, L.A. Cifuentes, B. Banrens, A. Charuey, and K.N. Dhanapala (2004), Energy management and global health. *Annual Review of Environmental Resources*, 29, pp. 383-419.

Faaij, A.P.C. (2006), "Bio-energy in Europe: changing technlogy choices", *Energy Policy*, 34, pp. 322-342.

Ha-Duong, M. and D. Keith (2003), "Carbon storage: the economic efficiency of storing CO2 in leaky reservoirs", *Clean Technology and Environmental Policy*, Special issue on Technologies for Sustainable Development, Vol. 5, No. 2/3, October.

Hoogwijk, M.M. (2003), "On the global and regional potential of renewable energy sources", PhD thesis.

Hotelling, H. (1931), "The Economics of Exhausitble Resources", *Journal of Political Economy*, Vol. 31, No. 2, pp. 137-175.

IAEA (2005), *Global Public Opinion on Nuclear Issues and the IAEA. Final Report from 18 Countries*, report prepared by Globescan ltd for the International Atomic Energy Agency.

IEA (2004a), *Biofuels for transport. An International perspective*, International Energy Agency, Paris, April.

IEA (2004b), *Energy security and climate change policy interactions: An assessment framework*, IEA information paper, International Energy Agency, Paris, December.

IEA (2005), *World Energy Outlook*, International Energy Agency, Paris.

IEA/OECD (2005a), *Prospects for hydrogen and fuel cells. Energy technology analysis*, International Energy Agency and Organisation for Economic Cooperation and Development, Paris

IEA/OECD (2005b), *Reducing Greenhouse gas emission. The potential of coal*, International Energy Agency and Organisation for Economic Cooperation and Development, Paris.

IIASA (International Institute for Applied Systems Analysis) (2005), *The GAINS Model for Greenhouse Gases - Version 1.0: Carbon Dioxide*, IIASA Interim Report IR-05-53 (available at www.iiasa.ac.at).

IPCC (2001a), *Climate change 2001: Mitigation*, contribution of Working Group III to the third assessment report of the intergovernmental Panel on Climate Change [Metz, B., O. Davidson, R. Swart, and J. Pan (eds.)], Cambridge: Cambridge University Press.

IPCC (2001b), *Climate change 2001: Impacts, Adaptation and vulnerability*, contribution of Working Group II to the third assessment report of the intergovernmental Panel on Climate Change, ISBN 0 521 01500 6, Cambridge, Cambridge University Press.

IPCC (2005), *Special report on CO_2 capture and storage*, summary for policymakers (www.ipcc.ch).

Jansen, J.C., W.G. van Arkel and M.G. Boots (2004), *Designing Indicators of long-term energy supply security*, ECN-C--04-007, Energy research Center of the Netherlands, Petten, NL.

Joosen, S. and K. Blok (2001), *Economic evaluation of carbon dioxide emission reduction in the household and services secotrs in the EU. Bottom-up analysis*. Final report (europa.eu.int/comm./environment/enveco).

Junginger, M. (2005), "Learning in renewable energy technologies", PhD thesis, Utrecht, The Netherlands.

Just, R.E., S. Netanyahu and L.J. Olsen (2005), "Depletion of natural resources, technological uncertainty, and the adoption of technological substitutes", *Resource and Energy Economics*, Vol. 27, No. 2, pp. 91-108.

Lako, P. (2004), *Coal-fired power technologies, Coal-fired power options on the brink of climate policies*, ECN report: ECN-C-04-076, Energy research Center of the Netherlands, Petten, NL.

Kok, M.T.J. and H.C. de Coninck (eds) (2004), *Beyond Climate, Options for broadening climate policy*, RIVM report 500019 001/2004.

Kroon, P., S.J.A. Bakker and H. de Wilde (2005), *NO$_x$ kleine bronnen. Update van de uitstoot in 2000 en 2010*, ECN report No. ECN-C--05-015, Energy research Center of the Netherlands, Petten, NL.

Little, I.M.D. and J.A. Mirrlees (1975), *Project Appraisal and Planning for Developing Countries*, New York: Basic Books.

MARKAL database

Markandya, A. and A. Hunt (2004), *Externalities of energy: extension of accounting framework and policy applications. The externalities of energy insecurity.*

Meadows, D.H., D.l. Meadows, J.Randers and W.W. Behrens (1972), *The Limits to Growth: A report for the club of Rome's project on the predicament of mankind*, New York, NY.

Menkveld (ed.) (2004), *Energietechnologieen in het kader van transitiebeleid. Factsheets*, ECN-C-04-20, Energy research Center of the Netherlands, Petten, NL.

Menkveld et al. (2005), *Het onbenut potentieel voor energiebesparing*, ECN-C--05-062, Energy research Center of the Netherlands, Petten, NL.

NEA/IEA (2005), *Projected Costs of Generating Electricity: 2005 Update"*, Organisation for Economic Cooperation and Development/International Energy Agency, Paris.

NERAC (2002), *A Technology Roadmap for Generation IV Nuclear Energy Systems*, US DOE Nuclear Energy Research Advisory Committee and the Generation IV International Forum, Washington, D.C., December.

Ogden, J.M., R.H. Williams and E.D. Larson (2004), Societal lifecycle costs of cars with alternative fuels/engines, *Energy policy*, 32, pp. 7-27.

OXERA (2002), *A Social Time Preference Rate for Use in Long-Term Discounting*, report to The Office of the Deputy Prime Minister, Department for Transport, and Department of the Environment, Food and Rural Affairs, Oxford, 17 December.

Palisade (2000), *@ Risk Standard Edition*, Version 4.0.5, Canada.

Porter M.E. and C. van der Linde (1995), "Toward a New Conception of the Environment-Competitiveness Relationship", 9, pp. 97-118.

Rabl, A. and J.V. Spadaro (2000), "Public health impacts of air pollution and implication for energy systems", *Annual review of Energy and the Environment*, 25, pp. 601-627.

Rabl, A., J.V. Spadaro and B. van der Zwaan (2005), "Uncertainty of Air Pollution Cost Estimates: To What Extent Does It Matter?", *Environmental Science and Policy*, 39, pp. 399-408.

Pearce, D.W. and R.K. Turner (1990), *Economics of natural resources and the environment*, Hertfortshire: Harvester Wheatsheaf.

Sijm, J.P.M., L.M. Brander and O.J. Kuik (2002), *Cost assessment of mitigation options the energy sector. Conceptual and methodological issues*, ECN report No ECN-C--02-040, Energy research Center of the Netherlands, Petten, NL.

Sathaye (2004), *Issues in conducting GHG mitigation assessments in developing countries*.

Smith (2001), *Potential for economic greenhouse gas reduction in coal-fired power generation*, IEA Clean Coal Centre report CCC/49, International Energy Agency, London.

Squire, L. and H.G. van der Tak (1975), *Economic Analysis of Projects*, Baltimore and London: Johns Hopkins University Press.

Thomas, S. (2001), *Critical Comments on the Use of the "Specific Costs of CO_2 reduction'as a Criterion for the Selection of Energy Resources*, Wuppertal Institute, 30 March.

Tinbergen, J. (1973), "Exhaustion and Technological Development: A macro-Dynamic Policy Model", *Zeitschrift für Nationalökonomie*, 33, pp. 213-234.

Tol, R.S.J. (2005), "The marginal damage costs of carbon dioxide emissions: An assessment of the uncertainties", *Energy Policy*, 33, pp. 2064–2074.

Torp, T.A. et al. (2004), "CO_2 underground storage costs as experienced at Sleipner and Weyburn", paper presented at GHGT-7, Vancouver, Canada, 6-9 September (http://uregina.ca/ghgt7/PDF/papers/peer/436.pdf).

Vito (2004), "Economische waardering van de milieu-impact van verzuring en vermesting", *Nieuwsbrief Milieu & Economie*, 2, p. 18.

Vuuren, D.P. van, J. Cofala, H.E. Eerens, R. Oostenrijk, C. Heyes, Z. Klimont, M.G.J. den Elzen and M. Amman (2006), "Exploring the ancillary benefits of the Kyoto Protocol for air pollution in Europe", *Energy Policy*, 34, pp. 444-460.

WADE (World Alliance for Decentralised Energy) (2005), *Projected cost of generating electricity (2005 update). WADE's response to the report of the International Energy Agency and the Nuclear Energy Agency*, August.

Wade, J. and A. Warren (2001), *Employment generation from energy efficiency programmes: enhancing political and social acceptability*, summer study proceedings (available at www.eceee.org).

Wakker, A., R. Egging, E. van Thuijl, X. van Tilburg, E. Deurwaarder, T. de Lange, G. Berndes and J. Hansson (2005), *Biofuel and Bioenergy implementation scenarios. Final report of VIEWLS WP5, modelling studies.*

Watkiss, P., C. Handley, R. Butterfield and T. Downing (2005), *The impacts and costs of climate change*, Final Report prepared for the European Commission DG Environment (http://www.europa.eu.int/comm/environment/climat/pdf/final_report2.pdf).

WHO (World Health Organisation) (2000), *Air quality guidelines*, 2nd edition (www.euro.who.int).

Wuppertal Institute fur Klimaschutz (2005), *Target 2020: Policies and measures to reduce greenhouse gases in the EU. A report on behalf of WWF European Policy Office, final report. Wuppertal* (available at www.panda.org).

Zwaan, B. van der (2005), "Will coal depart or will it continue to dominate global power production during the 21st century?", *Climate Policy*, 5, No. 4, pp. 441-450.

Annexes

A.1 Monte Carlo propagation analysis using @RISK

In order to be able to determine an uncertainty interval for the outcomes of the abatement cost calculations we carried out uncertainty propagation analyses for each CO_2 reduction option using @ Risk software (Palisade, 2000). This programme simulates the uncertainty in parameter assumptions (inputs) by varying them according to the uncertainty interval given by the user. Our outcomes are based on:

- Latin hypercube simulation (which uses stratified sampling techniques, resulting in convergence towards a sampled distribution in fewer samples than simple Monte Carlo simulation);

- 1000 iterations;

- Triangular probability distributions for the parameter values (see figure below for an example), based on the low, mean and high input values (as shown in Annex 4);

- No correlation between the inputs, which may not be completely true in practice. The outcomes can be seen as a conservative (optimistic) estimate of the uncertainty interval if, for example, two underlying factors having each a positive relationship with the dependent (i.e. the resulting cost variable) co-vary mutually in a negative (positive) way;

- For the outcomes, the low value represents the 2.5-percentile value and the high value the 97.5-percentile value, therefore the resulting bandwidth can be seen as the 95% confidence interval (based on the - low, mean, high - assumptions about the distribution of underlying factors).

Figure A.1 Triangular Input distribution for gas price escalator, with 2.5% value, mean value and 97.5% value 1.0, 2.0 and 4.0 %/yr respectively

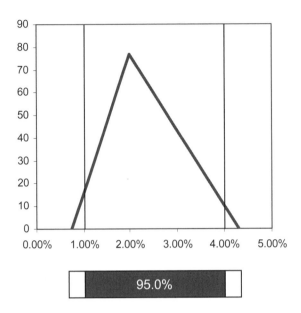

Figure A.2 Output distribution of abatement cost for CHP (replacing CCGT), with low, mean and high values in €/tCO₂ (for example purpose only).

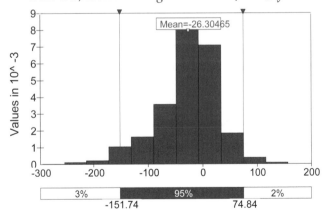

A.2 Pushing the sustainable technology frontier

Porter and Van der Linde (1995) provided some general policy prescriptions for environmental regulation to push the sustainable technology frontier which, put in a EU context, would seem worthy to be presented as a side-issue to this report. They recommend:

- *Maximum opportunity for innovation.* Define clear-targeted outcomes in a long-term time frame, leaving the choice of means, approach, and technology up to the private sector. Where possible, use market-based instruments;

- *Foster continuous improvement.* Refrain from locking in any particular technology as this does not provide stimulation in the direction of continued innovation. Notably, prescribing specific end-of-pipe technology should be avoided;

- *The regulatory process should leave as little room as possible for uncertainty at every stage.* Create new forums for settling regulatory issues that minimise deadweight loss creating litigation. Improve co-ordination of environmental regulation (i) between industry and regulators by early involvement of industry in setting standards, (ii) between regulators at different levels of member state administrations, (iii) among MS regulators and between EU regulators and overseas counterparts. Regarding sub-point (iii) Porter and Van der Linde (1995) suggest national, or rather in the present context the EU's regulations should be at least in sync but ideally slightly ahead of regulations in overseas countries. However, standards "too" far ahead would reduce or reverse early-mover advantages. Standards "too" different in character would lead industry to innovate in the wrong directions.

A.3   Technology development and exhaustible resources

Harold Hotelling formulated the so-called Hotelling rule that the (real) price of an exhaustible resource over time rises at a percentage equal to "the" discount rate (Hotelling, 1931). This can be explained as follows. The rule presumes perfect knowledge including perfect foresight. Point of departure is that at a certain point in future a "backstop technology" becomes available through which a substitute can be produced at a certain (currently non-competitive) price. Then it is optimal for the owners of the resource to set a current ask price for a unit of extracted resources and set the annual rate of extraction of the resource stock such that:

(i)   the economically extractable stock is depleted by the point in time the backstop technology becomes available

(ii)   during the transitional period the unit value of extracted resources equals the value of resources in the ground plus the extraction margin

(iii)   the capital tied up in resources in the ground will have a return comparable to the best alternative with similar risk profile as the resource extraction business.

Based on these type of assumptions the Hotelling rule can be understood intuitively and also be proved mathematically. A social discount rate including a margin for additional societal risk of the resource extraction business would seem in line with the perfect knowledge assumption. Yet as the business risk in the resource extraction business tends to be rather negatively than positively correlated to macro-economic growth cycles (oil and natural gas price hikes typically affecting economic growth in a negative way) no upward adjustment seems in order. Perfect knowledge regards the ultimately recoverable reserves situation and availability timing and cost of the backstop technology, while the social risk premium would not include the boom-bust cyclical short-term price oscillations around the long-term trend.

Allowance should be made for the impact of monopolistic market power of resource owners. For the monopolist, consistent with Hotelling's rule it is optimal that his marginal revenue, not the market price, rises at "the" discount rate. The initial market price would be higher, the rate of extraction lower, and the phase-out period of the exhaustible resource use retarded compared to (socially optimal) competitive market conditions (Devarajan & Fisher, 1981).

Hotelling's work triggered the interest of resource economists in the issue of the impact of unpredictable innovations. Just *et al* (2005) investigated the impact of uncertainty in the discovery date of superior backstop technologies, given an existing but not yet adopted backstop technology. Given certain model conditions, they prove that - prior to the discovery of a superior (more competitive) backstop technology than the existing one - the optimal discount rate should be higher than in the one of the Hotelling rule. Indeed, the optimal discount rate should contain an additional risk margin allowing for the possible event that a superior technology is discovered. It would occur to us that this is also consistent with a lower current ask price and a higher optimal current extraction rate and that, consequently, price trajectories would rise steeper than optimal in a Hotelling model world. We may infer that (chances, as perceived by market parties, notably resource owners, of) unforeseen discoveries of superior backstop technologies act as a countervailing factor to the exertion of market power. Well-designed R&D polices in oil and gas importing countries can stimulate targeted innovation offsets and thus strengthen this countervailing factor.

At the point in time of discovery of a highly superior technology than the pre-existing backstop technology, the trajectory of the resource price over time may shift in an upward direction. Discoveries of superior technology before an existing backstop technology gets adopted will speed up the time that the exhaustible resource will be totally phased out. Moreover, such discoveries will limit both the total amount of windfall royalty transfers to resource owners and will have a positive impact global welfare levels. The message to policy makers should be that a very high priority is warranted to design proper policy frameworks that foster acceleration of exhaustible-resource-saving innovations. The climate change issue has even enhanced the urgency for human kind to accelerate sustainability-enhancing technological development.

A.4 Assumptions used in the numerical example of social cost-benefit analysis of selected climate change mitigation options in the electricity generation sector

Assumptions for social CBA of selected options in the electricity sector

option	fuel		investment		fuel efficiency		O&M var	O&M fix	Load factor	lifetime	CEF	References
			2010 €/kW	2020 €/kW	2010 %	2020 %	€/MWh	€/kW	%	yr	tCO2/MWh	
PCC	coal	low	1000	980	40%	42%	2.6	19.6	75%	25	0.77	NEA/IEA, 2005
		mean	1100	1067	45%	48%	3.2	23.7	80%	30	0.85	Lako, 2004
		high	1200	1153	50%	54%	3.6	27.0	85%	35	0.85	Menkveld, 2004
CCGT	gas	low	450	441	50%	52%	1.4	14.7	70%	20	0.40	NEA/IEA, 2005
		mean	500	485	55%	58%	1.5	16.3	75%	25	0.37	Lako, 2004
		high	550	528	60%	64%	1.7	17.9	80%	30	0.34	Menkveld, 2004
wind on-shore		low	726	657			3.0	28.6	23%	10	0	NEA/IEA, 2005
		mean	887	700			5.4	35.7	29%	15	0	CPB/ECN, 2005
		high	1026	682			7.2	42.8	34%	20	0	Menkveld, 2004
IGCC	coal	low	1200	1141	48%	51%	2.5	42.0	80%	25	0.57	NEA/IEA, 2005
		mean	1467	1304	52%	56%	3.1	52.5	83%	30	0.67	Lako, 2004
		high	1700	1389	56%	61%	3.7	63.0	85%	35	0.67	
Nuclear LWR/EPR	uranium	low	1330	1304	32%	34%	3.4	25.7	87%	30	0.00	NEA/IEA, 2005
		mean	1900	1844	36%	39%	3.9	29.4	90%	40	0.05	Menkveld, 2004
		high	2470	2373	40%	44%	4.5	34.2	93%	50	0.10	
Biomass co-firing	biomass	low	472	449	33%	35%	12		70%	15	0.00	NEA/IEA, 2005
		mean	590	525	37%	40%	18.5		75%	20	0.10	CPB/ECN, 2005
		high	708	578	41%	45%	25		80%	25	0.20	ECN 2005
CHP (CCGT)	gas	low	450	428	37%	39%	4.7		70%	20	0.27	NEA/IEA, 2005
		mean	633	563	40%	43%	6.6		75%	25	0.30	Donkelaar et al, 200
		high	800	654	43%	47%	8.4		80%	30	0.33	Menkveld, 2004

For CHP, the thermal efficiencies are 39%, 41% and 43% and the efficiency of the reference heat generation plant 85, 87 and 90% in the 2.5%, average and 97.5% cases respectively.

Unless specifically stated otherwise we do not explicitly account for possible future policy intensification for the reason that this is a generic and preliminary assessment of distinct policies.

It is acknowledged that in concrete, location-specific cases of proposed GHG reduction activities explicit allowance should be made for the evolution of the policy framework during the envisaged activity time horizon. Moreover, ideally interactions between policies and the portfolio of climate change abatement options should be modelled as a major factor driving "endogenous learning" with feedback mechanisms for mutual market deployment interactions between climate change mitigation options.

A.5   Time trajectories of oil and gas risk premiums

This annex visualises the time trajectories of the oil and gas prices in €/GJ, including the risk premiums set out in Chapter 5. The lower lines in both graphs represent the low oil/gas price case including the risk premium. The middle line (triangle data points) is the medium price scenario including the mean premium. The upper line represents the high price scenario in which it is presumed that the externality of energy security of supply is properly represented, i.e. the risk premium is equal to zero.

It can be observed that until year 2017, the medium oil price scenario including risk premium is equal to the high price scenario. Until that year the condition that the assumed oil price plus risk premium should not exceed the (assumed) oil price upper bound, is binding. For gas, this holds true until 2024, given price and premium assumptions specified in the main text.

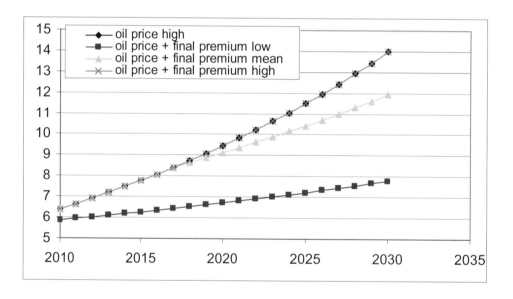

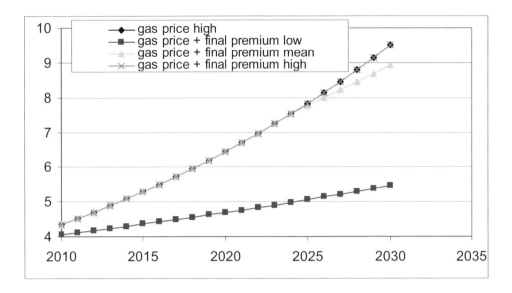

GLOSSARY OF ABBREVIATIONS AND TECHNICAL TERMS

A1B	IPCC scenario
AEO	Annual Energy Outlook (by the US Department of Energy)
AP	Air pollution
AQ	Air quality
B2	IPCC scenario
BE	Built Environment
BOE	Barrel of oil equivalent
CBA	Cost-benefit analysis
CCGT	Combined cycle gas turbine
CCS	Carbon Capture and Storage. Technologies to capture and store CO_2 in geological structures
CCS	CO_2 capture and storage
CDM	Clean Development Mechanism (Kyoto flexible mechanism)
CEF	CO_2 emission factor
CEPS	Centre for European Policy Studies
CHP	Combined heat and power (co-generation)
CHP	Combined Heat and Power (co-generation), which has a conversion efficiency of 70% or more
CO_2	Carbon dioxide, the main greenhouse gas (GHG) covered in the Kyoto Protocol
DoE	Department of Energy
DSM	Demand-side management (energy-efficiency)
EBRD	European Bank for Reconstruction and Development
EC	European Communities, referring to the economic competencies of the European Union
ECCP	European Climate Change Programme, the European Commission's programme to consult with stakeholders on climate change
ECN	Energy research Centre of the Netherlands
EEA	European Environment Agency
EIA	Energy Information Administration
EIT	Economies-in-transition (among others new and accession member states)
EOR	Enhanced Oil Recovery
EPB Directive	Directive (2002/91/3C) on energy performance of buildings
EPBD	European Performance of Buildings Directive
ESS	Energy Security of Supply
EU ETS	EU Emissions Trading Scheme, covering CO_2 emissions from industry and the power sector

EU	European Union (see also EC)
EURIMA	European Insulation Manufacturers Association
European Council	Regular meetings of the heads of all EU governments to discuss and set out the strategic direction of the EU
G8	Regular summit of the heads of the eight most important economies in the world
GEM	General Equilibrium Model
GHG	Greenhouse gas
GHG	Greenhouse gas, usually referring to one of the six gases covered by the Kyoto Protocol: carbon dioxide (CO_2), methane (CH_4), nitrous oxide (N_2O), hydrofluorocarbons (HFCs), perfluorocarbons (PFCs) and sulphur hexafluoride (SF_6).
GJ	Gigajoule (10^9 joule primary energy)
Greenhouse effect	The earth has a natural temperature control system. Certain atmospheric gases are critical to this system and are known as greenhouse gases. On average, about one third of the solar radiation that hits the earth is reflected back to space. Of the remainder, some is absorbed by the atmosphere but most is absorbed by the land and oceans. The earth's surface becomes warm and as a result emits infrared radiation. The greenhouse gases trap the infrared radiation, thus warming the atmosphere. Naturally occurring greenhouse gases include water vapour, carbon dioxide, ozone, methane and nitrous oxide, and together create a natural greenhouse effect. Human activities are causing greenhouse gas levels in the atmosphere to increase and this has occurred to such a level as to bring about climate change.
GtC	Gigatonne of carbon (1 Gt = 1,000 Mt)
$GtCO_2$	Gigatonne of carbon dioxide
IAEA	International Atomic Energy Agency
IEA	International Energy Agency
IEA	International Energy Agency
IGCC	Integrated Coal Gasification Combined Cycle
IGCC	Integrated gasification combined cycle (power plant)
IIASA	International Institute for Applied Systems Analysis
IPCC	Intergovernmental Panel for Climate Change, a scientific body created by the UN, generally assumed to be the most authoritative source on climate change science, which operates on the basis of peer review
Kyoto Protocol	1997 Protocol under the UNFCCC to reduce GHG emissions globally. It entered into force on 16 February 2005 and will cover the period from 2008-2012; After 2012, a new framework or protocol will be needed. See "post-2012 framework"
LWR	Light water reactor
mb/d	Million of barrels per day (measurement)
MNP	Netherlands Environmental Assessment Agency
Mt	Million of tonnes. One Mt of CO_2 in the atmosphere is equivalent to 0.3 Mt carbon
$MtCO_2e$	Millions of tonnes of carbon dioxide equivalent, the most commonly used way to express quantities of GHGs

n.e.s.	not elsewhere specified
NEA	Nuclear Energy Agency
NGO	Non-Governmental Organisation
NO_x	Oxides of nitrogen (acidifying substances and PM precursor)
O&M	Operating and maintenance (cost)
PCC	Pulverised coal combustion (power plant)
PM	Particulate matter
Post-2012 framework	Describes the – yet to be established – global framework beyond 2012 to reduce GHG emissions, when the Kyoto Protocol expires
ppm/ppmv	Parts per million/parts per million volume, the most commonly used way to express quantifies of GHG concentrations in the atmosphere. Usually expressed in CO_2-equivalent whose value is established on the basis of the Global Warming Potential (GWP) for each GHG
PV	Photo-voltaic
PWh	Peta-watthour (10^{15} Wh)
R&D	Research and development, sometimes also called RTD, research and technological development or RD &D, research, development and deployment
REACH	A proposed and soon to be adopted registration and authorisation procedures for chemical substances in the EU
RES-E	Renewable energy sources for electricity production
SO_2	Sulphur dioxide (idem)
TWh	Terawatthour (10^{12} Wh)
UNFCCC	United Nations Framework Convention on Climate Change, agreed at the UN Conference on Environment and Development (Rio de Janeiro, 1992). The ultimate objective of the UNFCCC is to stabilise GHG emissions at a level that would prevent dangerous anthropogenic interference with the climate system. The most important climate agreement negotiated in the UNFCCC so far is the Kyoto Protocol
USGS	US Geological Survey
VOC	Volatile organic compounds
VSL	Value of statistical life
WBCSD	World Business Council for Sustainable Development
WEO	World Energy Outlook
WTP	Willingness to pay
YOLL	Years of life lost